Slim Pickins

Tom Policano

First Printing: 2014

ISBN: 978-0-9915057-1-5

IFOTOZ INC Publications

P. O. Box 20047 Rochester, NY 14602

http://ifotoz.com

Book Design -Ken Hoffmann
Copy Editor - Jonathan Secon

Slim Pickins

Foreword

Photographer Tom Policano, in his exquisite photo documentary Slim Pickins, has captured the faces and stories of humanity on the move. In 1970-71 Tom spent nearly two years living and working with migrant farmworkers who traveled up and down the East Coast in search of work as manual laborers in the fruit and vegetable agricultural industry. Was it a coincidence or maybe a calling that, following graduation from college, Tom volunteered to find and transport migrant workers in need of health services. Tom who had a camera but not a car? The organizer of the migrant health clinic, John Radebaugh M.D., a pediatrician and professor at the University of Rochester Medical Center, came to the rescue by arranging for Tom to use an old Rambler station wagon. What followed is presented in this book.

During World War II, growers of the fresh and processed food Americans enjoyed each and every day struggled to find the local labor needed to harvest and process their crops. The solution became a hodgepodge network of long-distance recruitment, transportation and temporary housing, with minimal regulatory oversight, of seasonally employed agricultural workers from Florida and other Southern states. Seventy-five years later, this practice has become institutionalized and remains the primary source of seasonal agricultural labor in America. Men, women, and their children were recruited and employed by farm labor contractors (crew bosses), who controlled everything from transportation to the company store. Workers traveled in rickety buses and cars, lived in dilapidated shanties and worked for subsistence wages. They endured the unrelenting and unpredictable environment that engulfs the harvest of food destined for our dinner tables. As I describe migrant life of the 70's, I must also report that the living and working conditions for today's migrant workers are not measurably better. Today, the faces of farmworkers are predominantly Mexican, South American and Jamaican, many of whom are forced to live in the shadows of a broken immigration system. Their struggle to survive is as predictable as it was when Tom photographed more than 40 years ago. Throughout the years, there have been countless protests by farmworkers and their advocates to demand that farm labor be as respected and valued as any other essential workforce. A combination of governmental funding, philanthropic contributions and passionate volunteer advocates have created many important services designed to intervene and support thousands of workers earning wages below the poverty line. Then there are programs that provide education, healthcare, job training, legal services and safe and decent housing that have made life for farmworkers a little more tolerable. These services, however, are but public subsidies to the agricultural industry and to us as consumers. Migrant and seasonal farmworkers in 2014, through heavy lobbying by agribusiness protagonists, are still excluded from national and state laws and regulations that protect the wages, hours, safety and working conditions of our nation's workforce.

Those who view the photographs in Slim Pickins will have many reactions dependent upon their own life story, political persuasion, and knowledge of the sophisticated harvesting system required to deliver to market our fresh, ripe, but not too ripe, fruits and vegetables. It is my guess, though, that as you enjoy your next apple, orange, peach, cherry, onion or salad, you will think about the human toll it took to bring that delicious life-giving produce to your table.

These photos depict the valor of those relegated to the labor sector of our food system that is designed to sustain a cost-effective ("cheap") abundant food supply. The system requires a labor force that is willing to be uprooted from their families and communities. A labor force willing to eke out only enough money to sustain life and limb. I have spent my entire career working with and advocating on behalf of migrant and seasonal farmworkers. My exposure to migrant farmworkers began in the summer of 1967, when I was employed by a group of United Methodist churches to provide outreach services to migrant farmworkers living in their communities. Throughout my career, I have provided leadership in several organizations established to serve migrant and seasonal farmworkers. I met Tom soon after he had left the migrant stream, and just after he had created a multi-image presentation of the photographs depicted in Slim Pickins. The presentation of his photos, each flowing into the other with a music track of blues music, was an extremely powerful and emotional experience. The audience had the opportunity to experience life in the migrant stream through the lens of Tom's camera. Tom was not an outsider taking shots of his subjects; rather, he worked side-by-side with, and got to know people intimately. He developed trusting relationships built on mutual respect and shared experiences.

These photographs are frozen in time and represent a cross-section of workers and their families that happened to be in the migrant stream in 1970-71. Tom did not start out expecting to follow the stream for two years. Encouraged by Beaumont Newhall, then curator of the George Eastman House of Photography in Rochester, NY, he applied for and received a New York State Council on the Arts grant for his early work. This money enabled Tom to maintain the car he was given and pay for travel and supplies to continue his work. Then, to postpone his student loan payments from engineering school, Tom enrolled in the first class of the Visual Studies Workshop, an extension program of SUNY Buffalo in Rochester, NY created by Nathan Lyons. Slim Pickins became the subject of his thesis presentation for a Masters in Fine Arts. In 1976, Tom was hired as a photographer at Rochester Institute of Technology, and two years later, became faculty. In 2014, he is still teaching photography at RIT. Throughout his life, photography has been Tom's passion.

In Slim Pickins you will meet Roberta Frasier. I have had a photograph of Roberta in my office since Tom gave it to me in 1973. I did not know her name or story until just recently, when Tom shared what he remembered from working with her. Roberta, a single young woman in need of a job, decided to travel and live with a crew of workers. She toiled alongside men and women to harvest whatever crop her crew boss negotiated with the grower. She stands tall, strong and defiant in the middle of a field harvesting life-giving food for others. Her look is defiant, her eyes piercing into the eyes of the beholder. She is resilient in her beauty and resolute in her determination to make something of herself. She gives me courage each and every day to take on the principalities and powers that have the opportunity and responsibility to ensure human beings like Roberta are treated with the respect they deserve. I do wonder where Roberta is today?

I am thankful that Tom has documented a portion of his life journey in Slim Pickins. He has created a beautifully organized book that captures what he saw. In the book is an image of a hugh stack of government reports from hearings conducted in order to find ways to improve farmworker living and working conditions. Though *Migrant and Seasonal Farmworker Powerlessness* is the title for the reports in this photo, I see something different. In Tom's photos, I see determination, perseverance, camaraderie, and a commitment to survive and even thrive. Tom's humanity, sensitivity and intensity has put all of this in full view. Slim Pickins is a treasure for anyone who is interested in meeting those who harvest our fruits and vegetables.

Stuart J Mitchell
President/CEO
PathStone Corporation

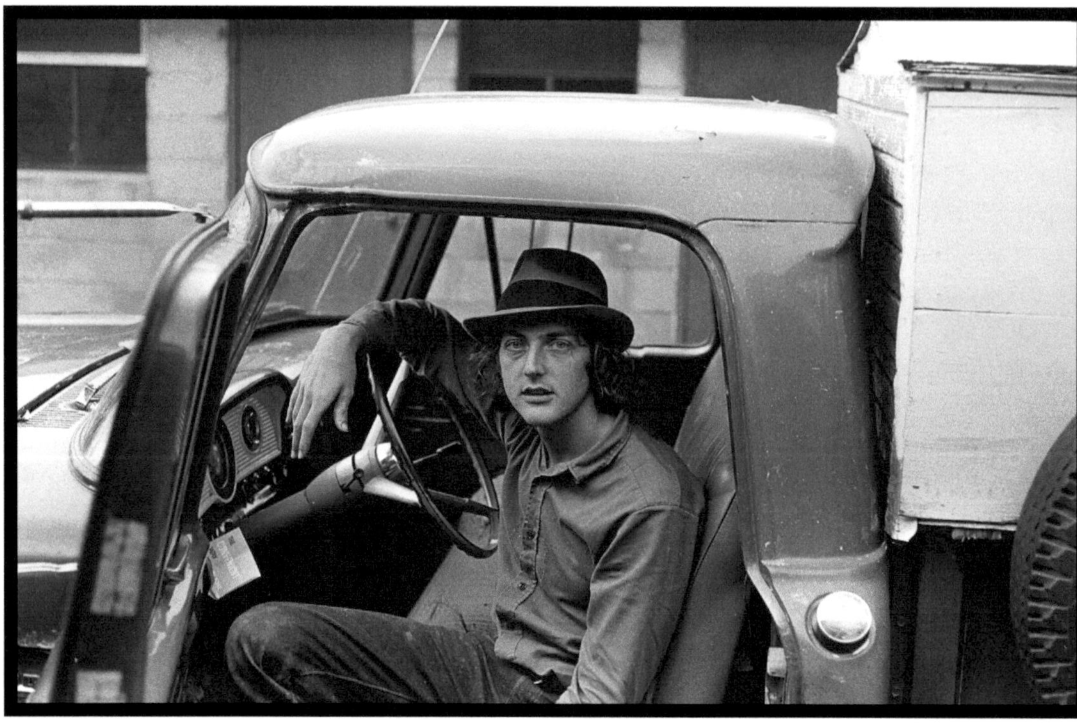

Introduction

After graduating from the University of Rochester's School of Electrical Engineering in 1969, I decided to postpone beginning my career in computer programming. It was now ten years after *Harvest of Shame*, the 1960 television documentary depicting the plight of the American agricultural farm worker presented by broadcast journalist Edward R. Murrow on CBS. My idea was to test my potential as a professional photographer by doing a follow-up documentary on the East Coast Migrant Stream. To begin, I became a volunteer driver helping migrants get to a free migrant health clinic run by Dr. John Radebaugh. I was lent an old Rambler station wagon to get around and ended up spending two years living and working in the East Coast Migrant Stream. Slim Pickins is my chronicle of that time. This journey changed my life.

The book consists primarily of photographic images detailing my impressions of seasonal farm workers and glimpses of their lives in the fields and orchards, in camps, at home and in the news. Interspersed are sections of text from conversations with Reverend Peterson and Johnny Wilcox, conversational snippets remembered, and a brief section from my diary in which I document attempts to organize East Coast migrant farm workers.

Migrant workers are an interstate agricultural labor force. In 1971, the East Coast Migrant Stream consisted of about 60,000 workers who were employed in Florida most of the year. About 12,000 traveled to New York during the summer season.

At this time the total number of interstate workers entering New York State was decreasing. About 76% of the workers were blacks from the southeastern section of the U.S., mainly Florida; about 16% were Puerto Ricans, either directly from the Island or from Florida; and the balance were primarily Mexican-Americans from Texas and Florida. The southern migrant workers consist of small families and individuals, and crews, composed of a few families and a predominance of unattached males.

North — At Work

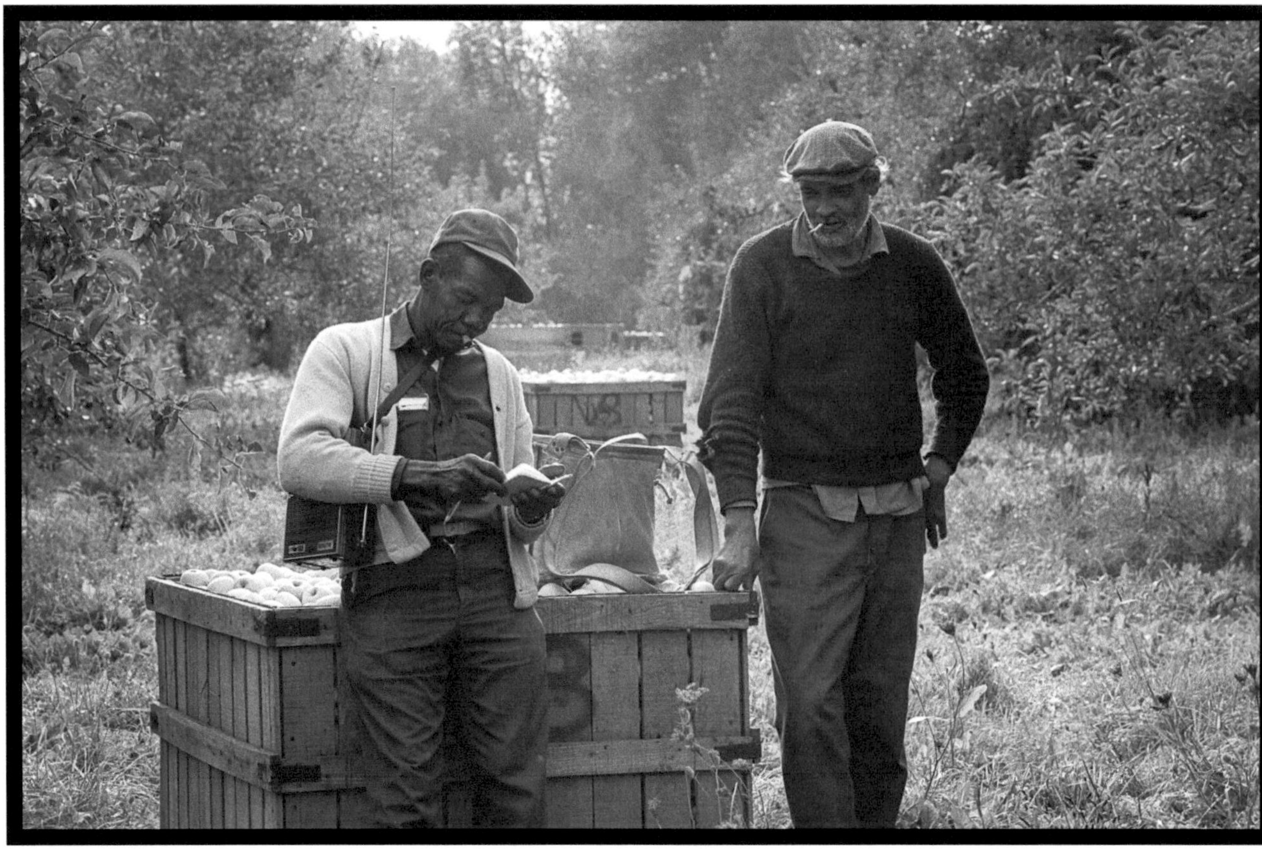

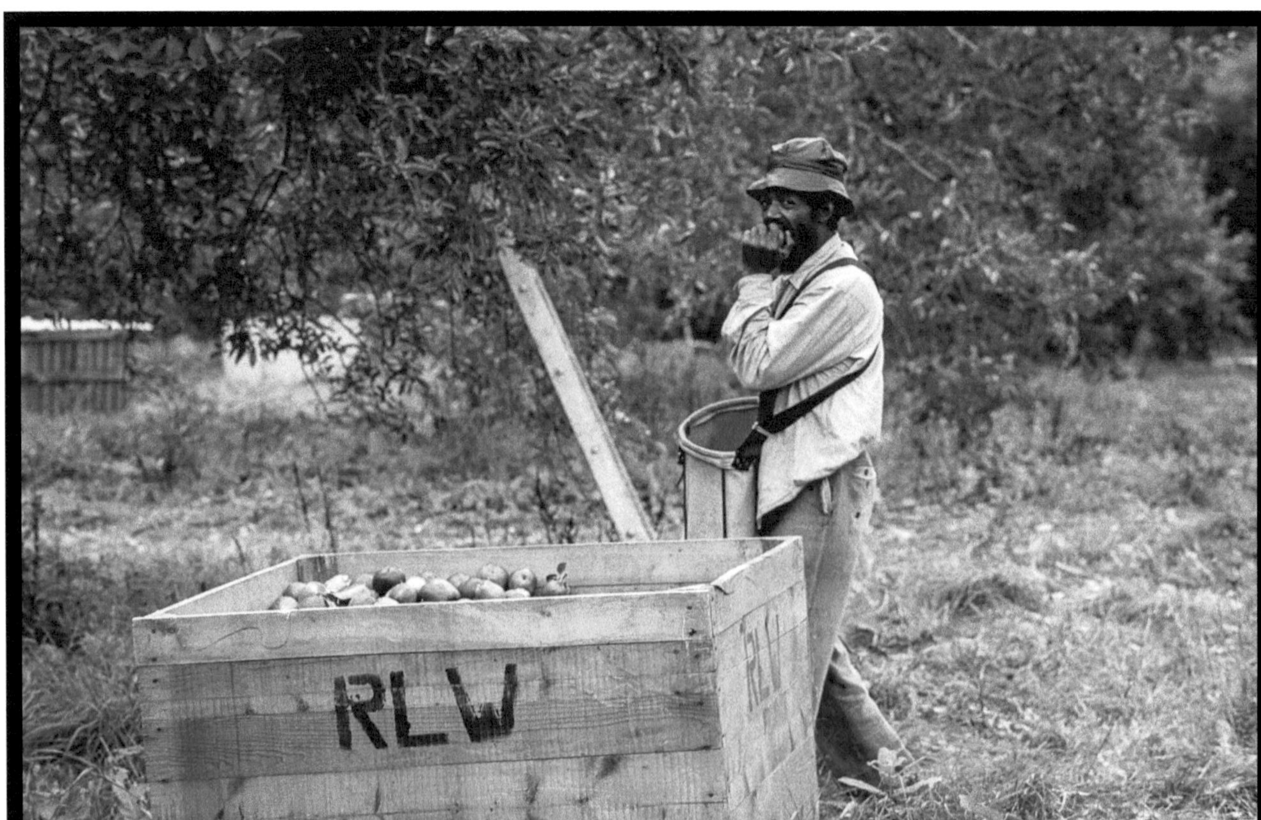
RLW
RLW

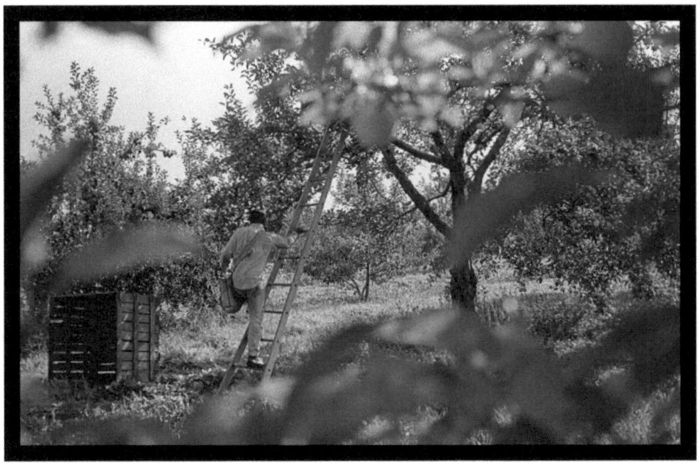

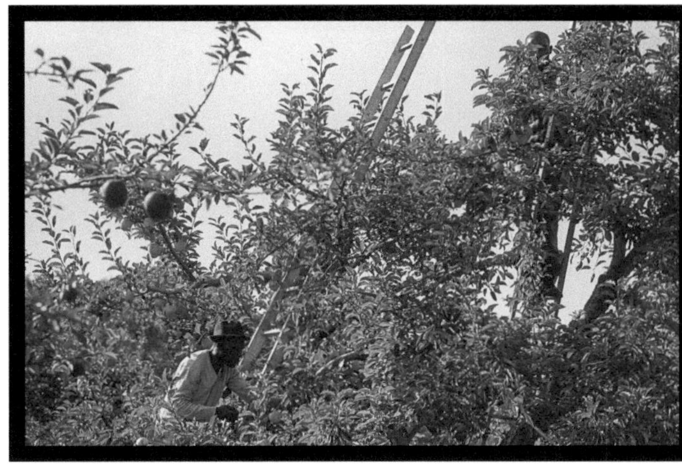

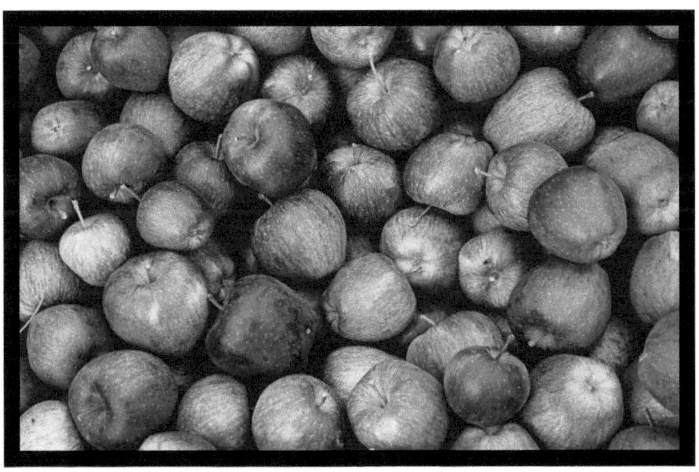

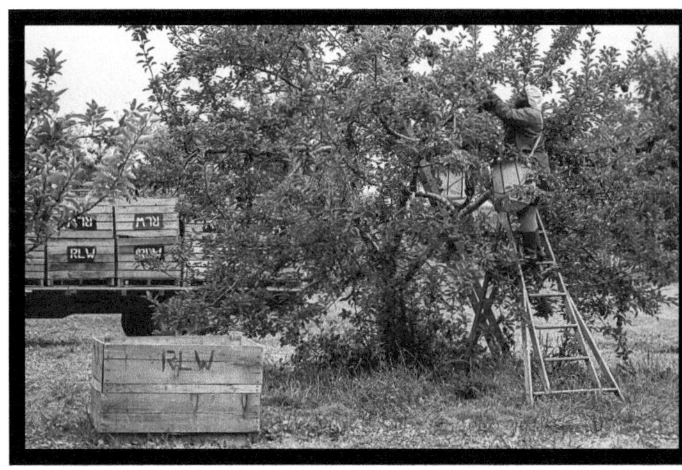
RLW

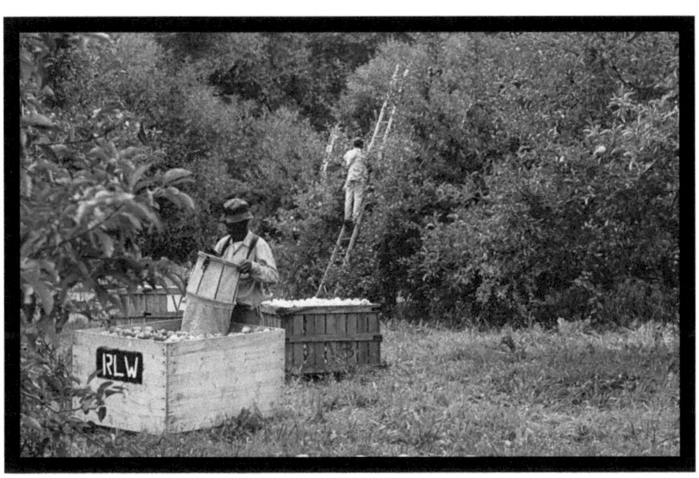
RLW

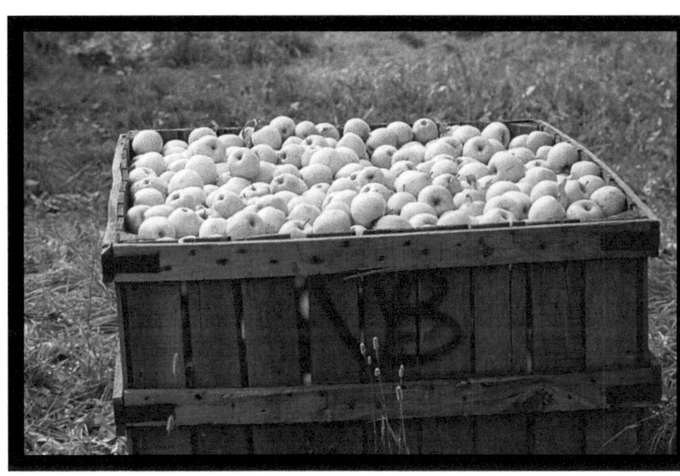
NB

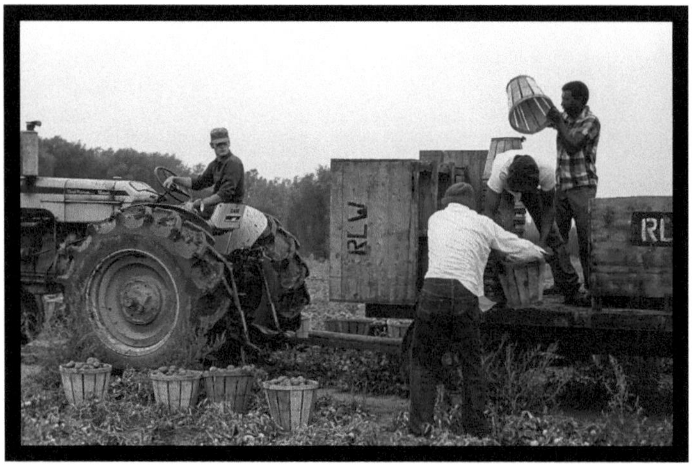
RLW
RL

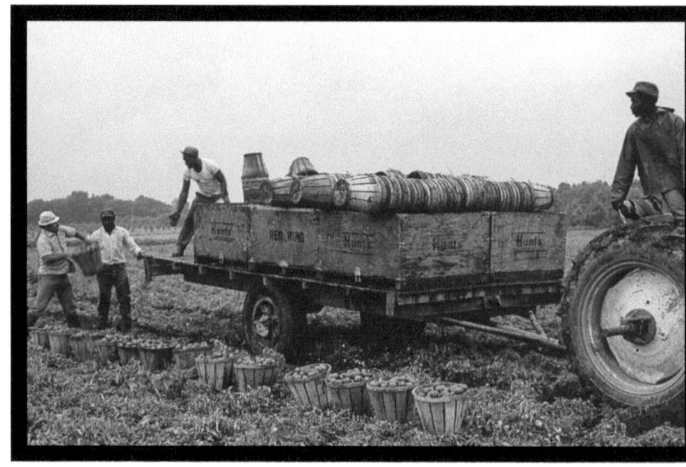

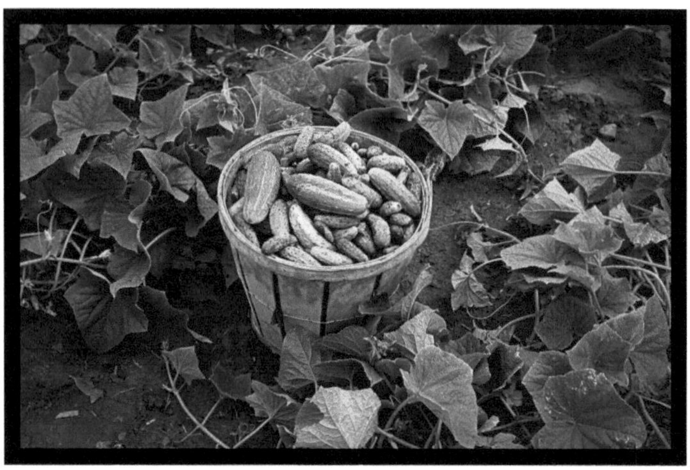

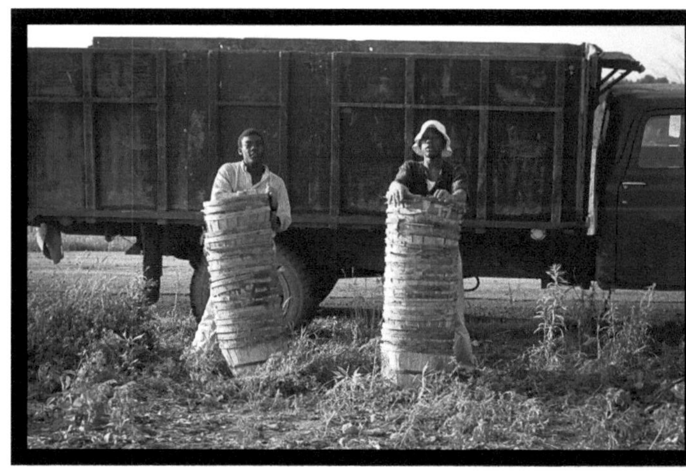

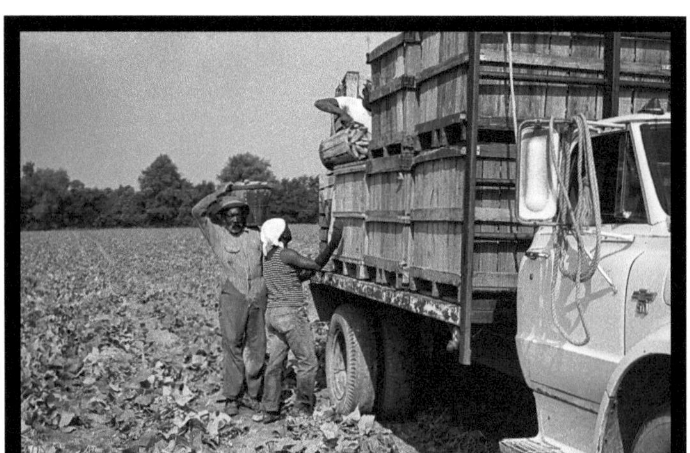

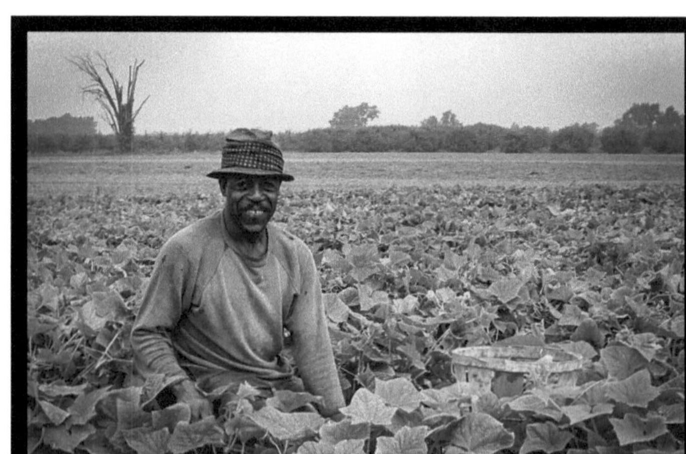

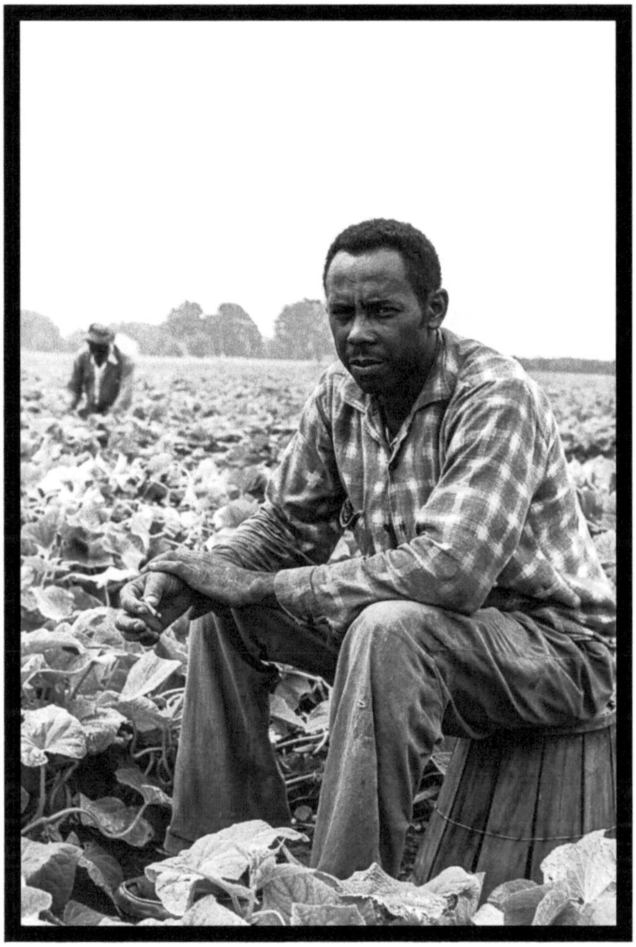

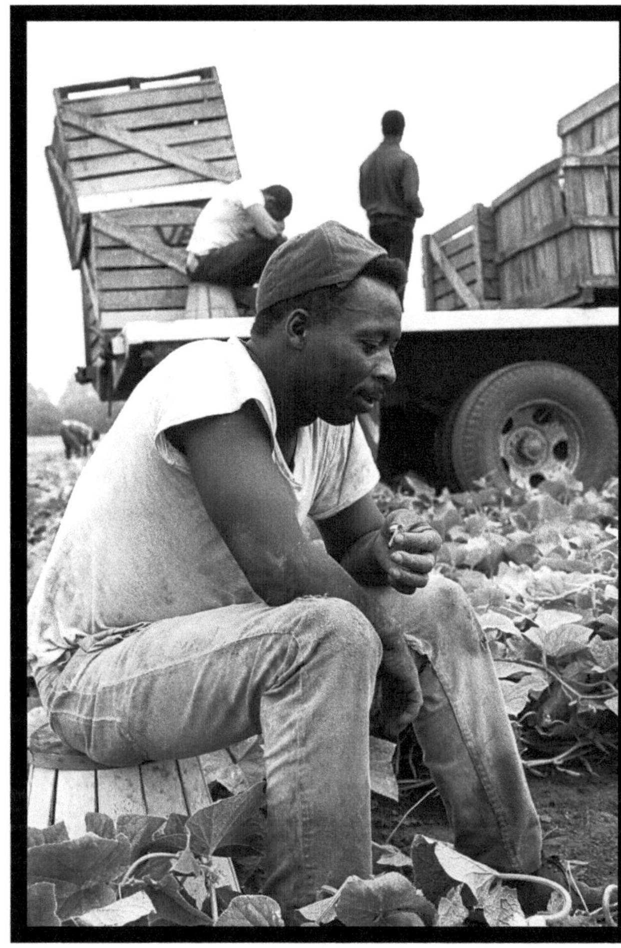

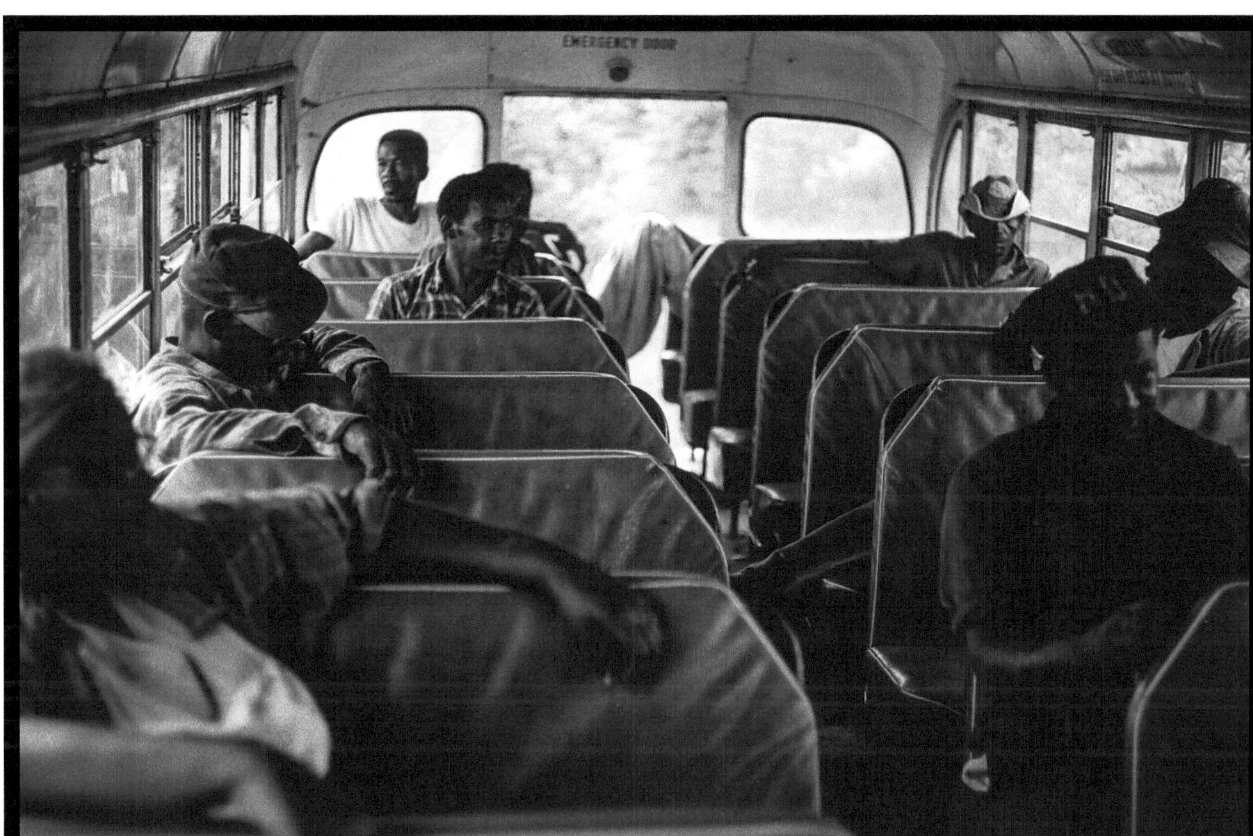

Tom Policano

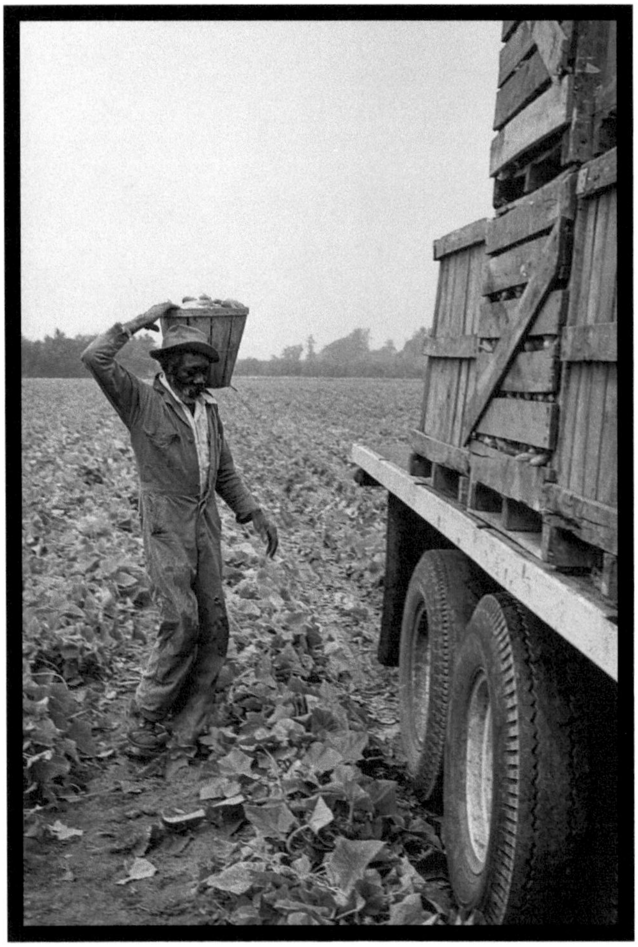

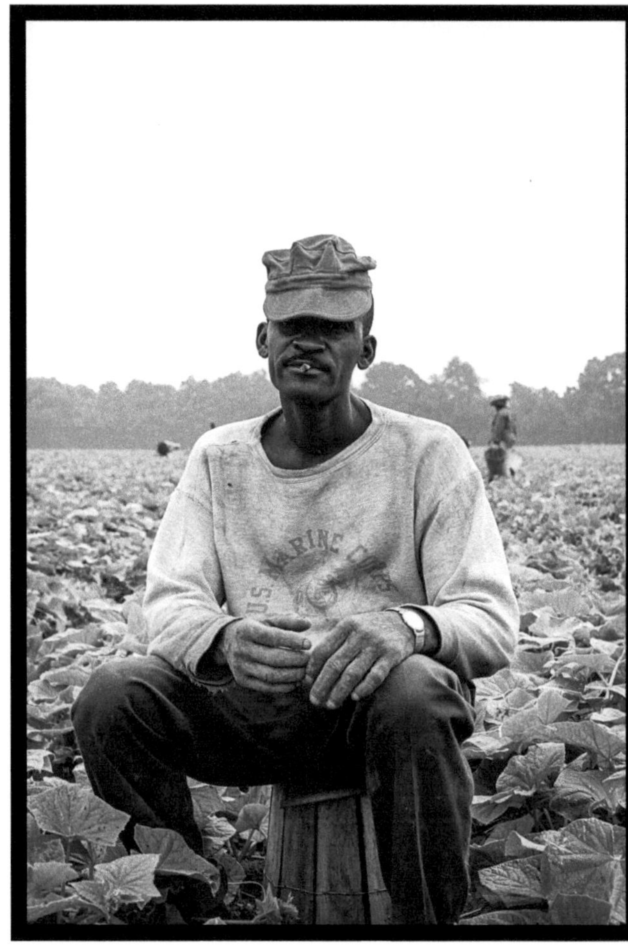

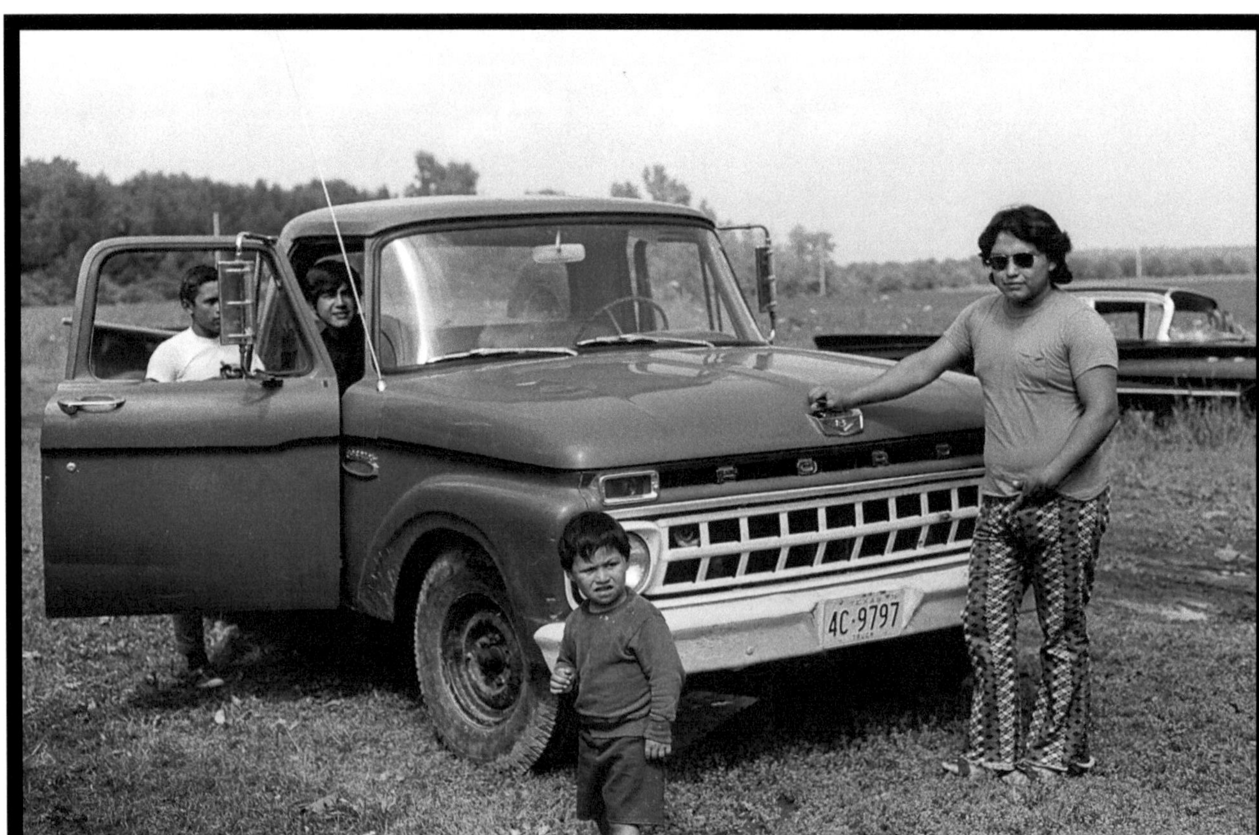
4C·9797

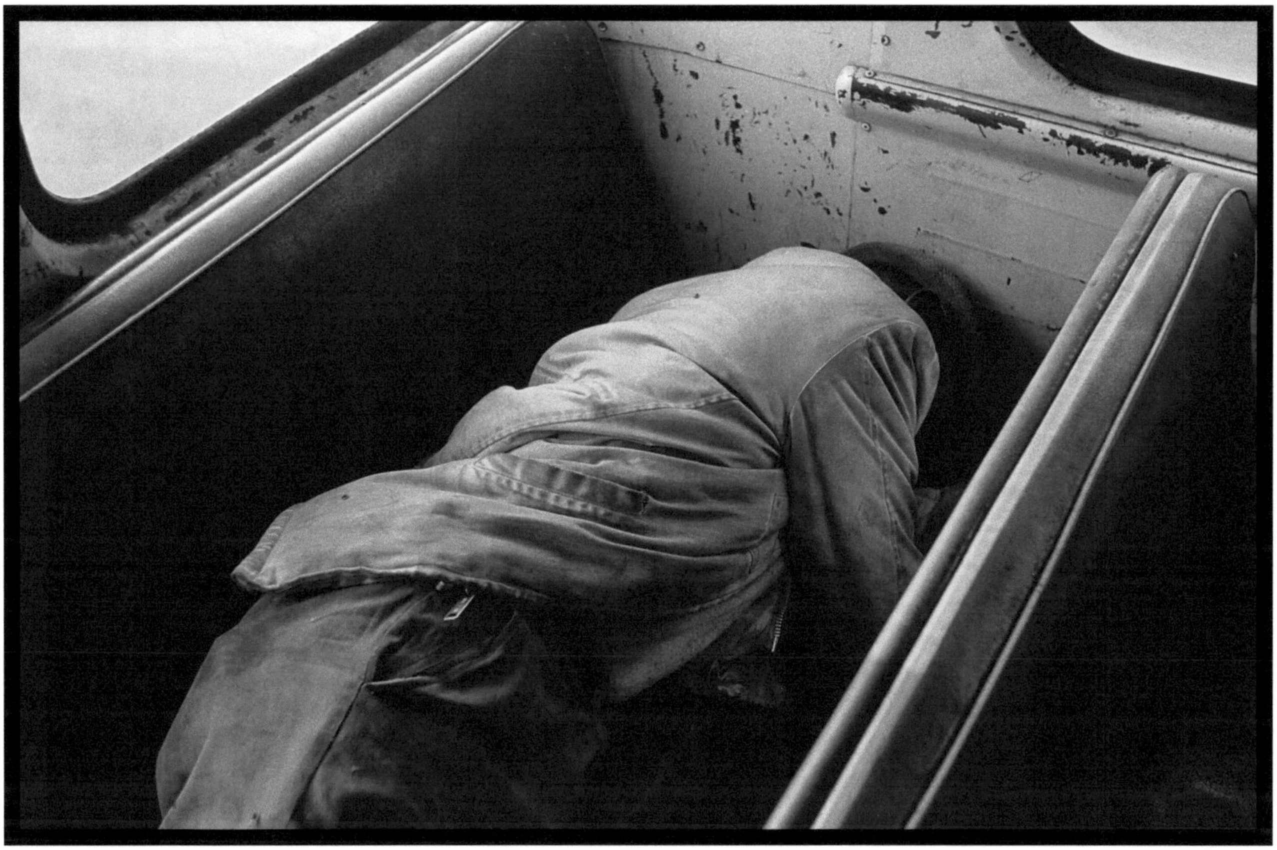

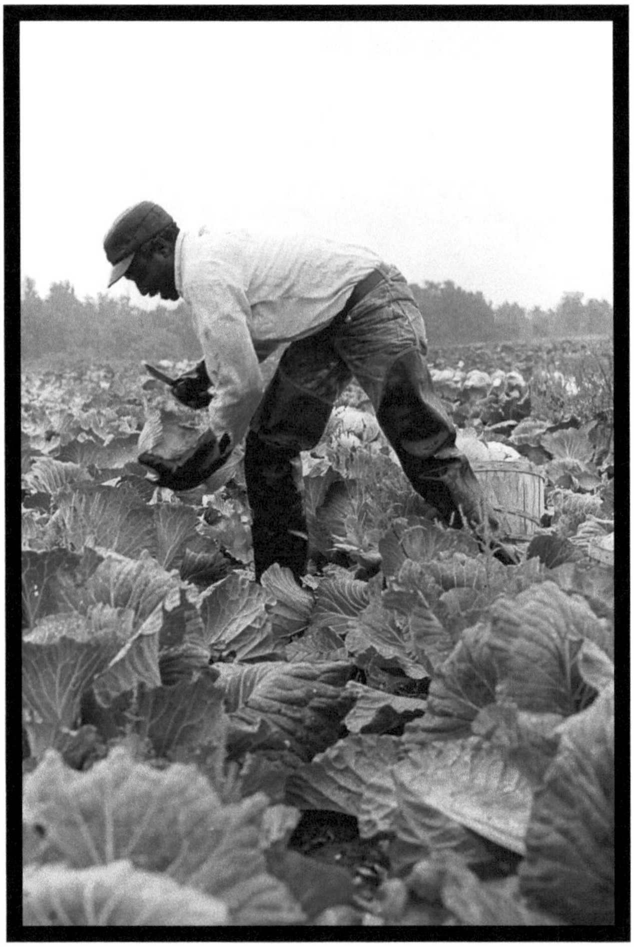

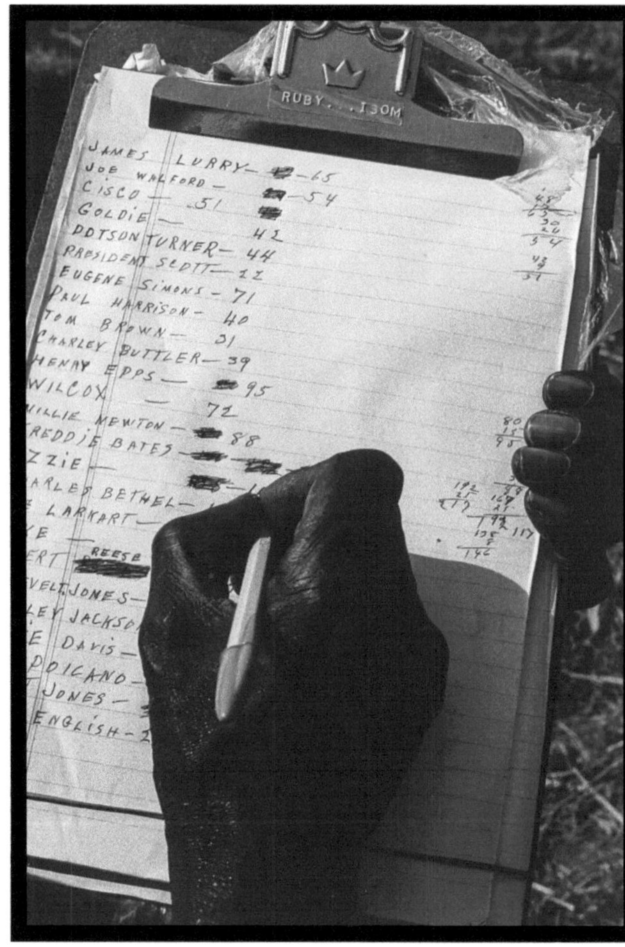
JAMES LURRY — 65
JOE WALFORD — 54
CISCO — 51
GOLDIE — 42
DOTSON TURNER — 44
PRESIDENT SCOTT — 22
EUGENE SIMONS — 71
PAUL HARRISON — 40
TOM BROWN — 31
CHARLEY BUTTLER — 39
HENRY EPPS
WILCOX — 95

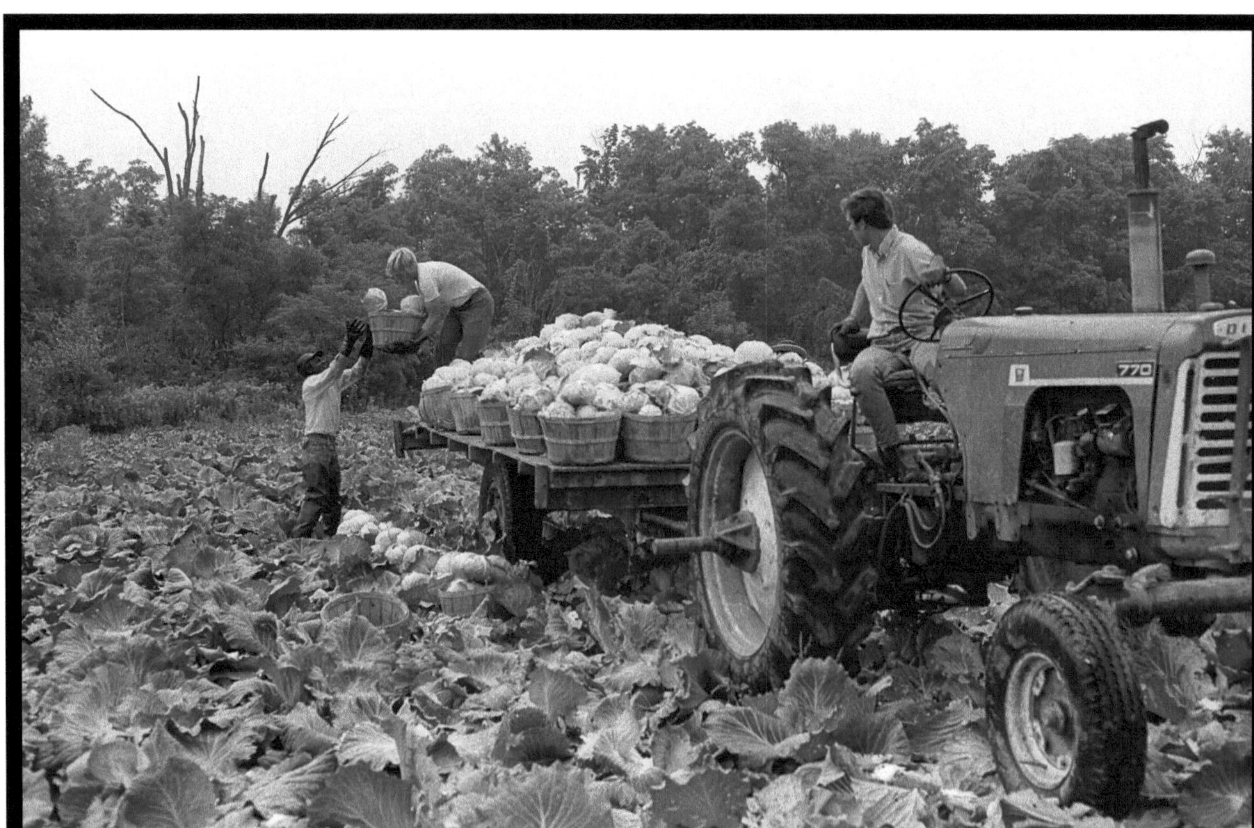

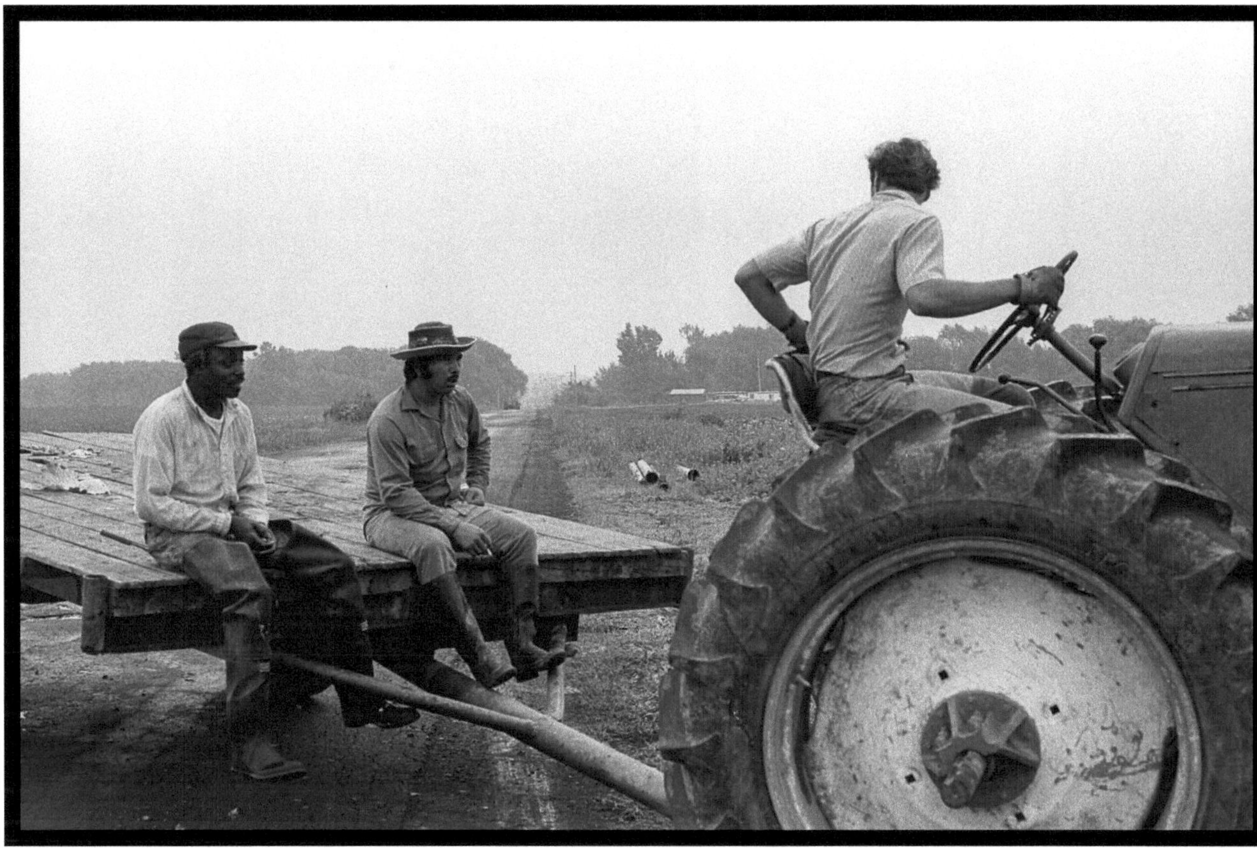

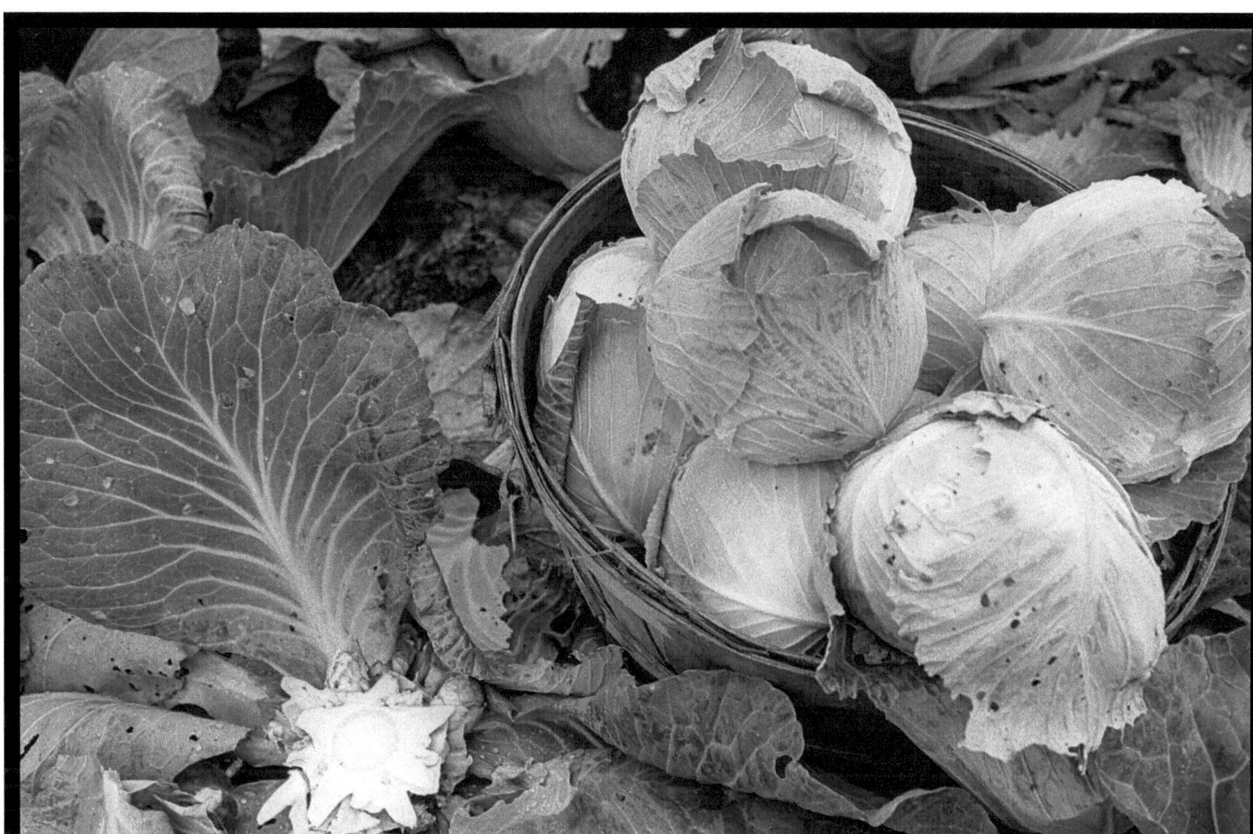

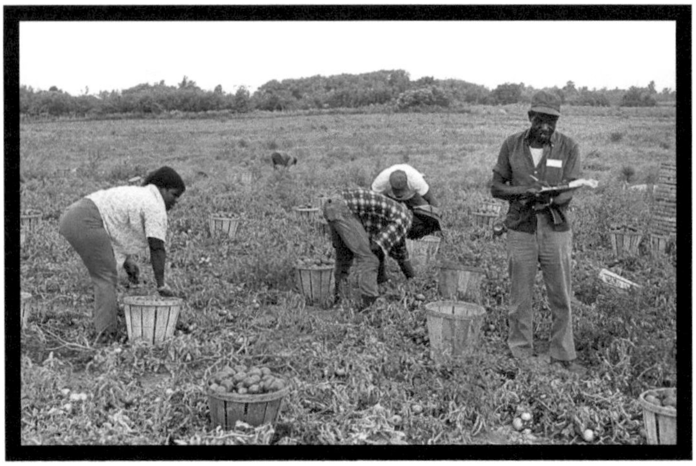
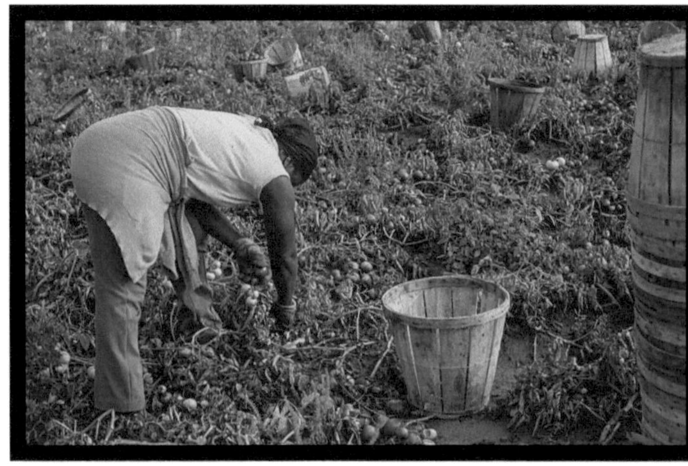
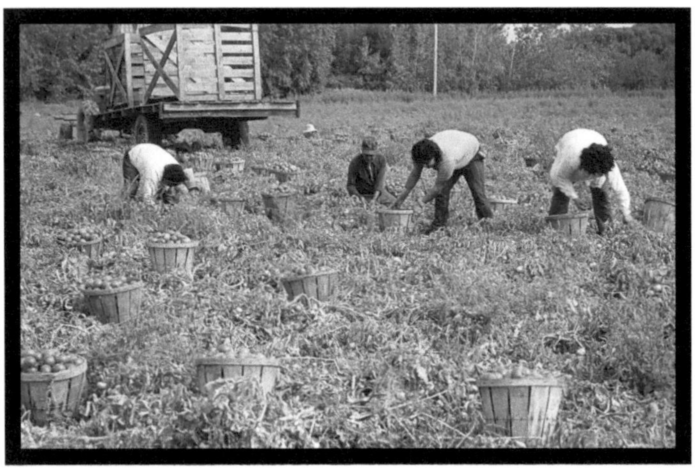
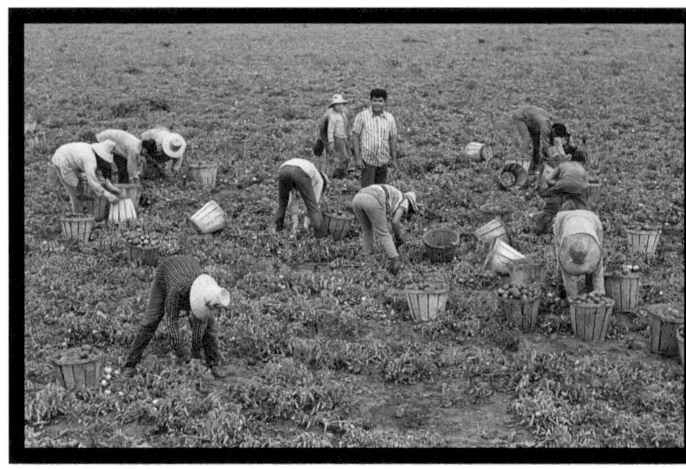
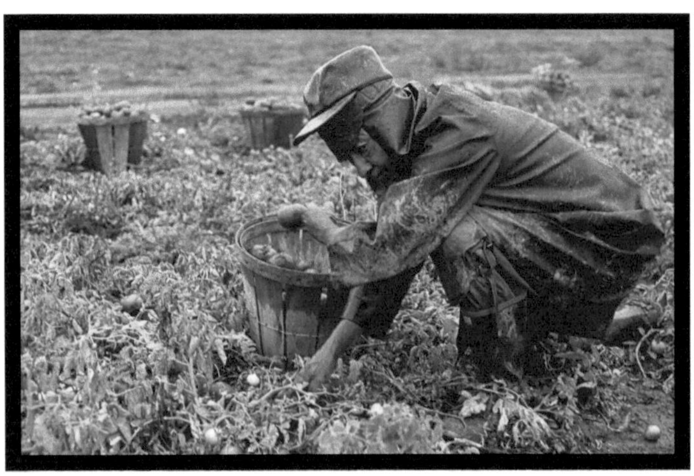
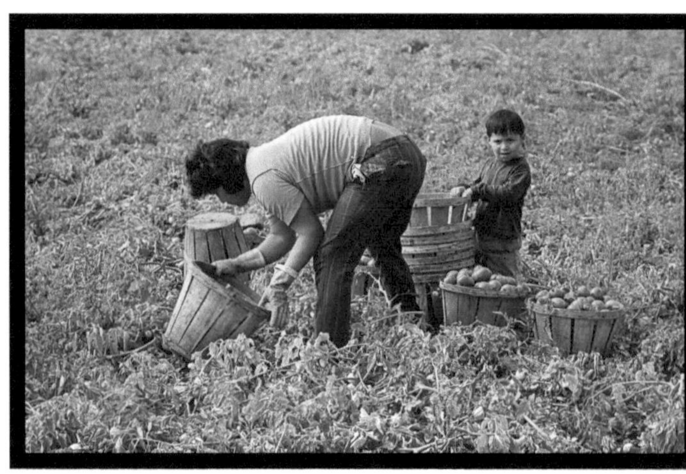

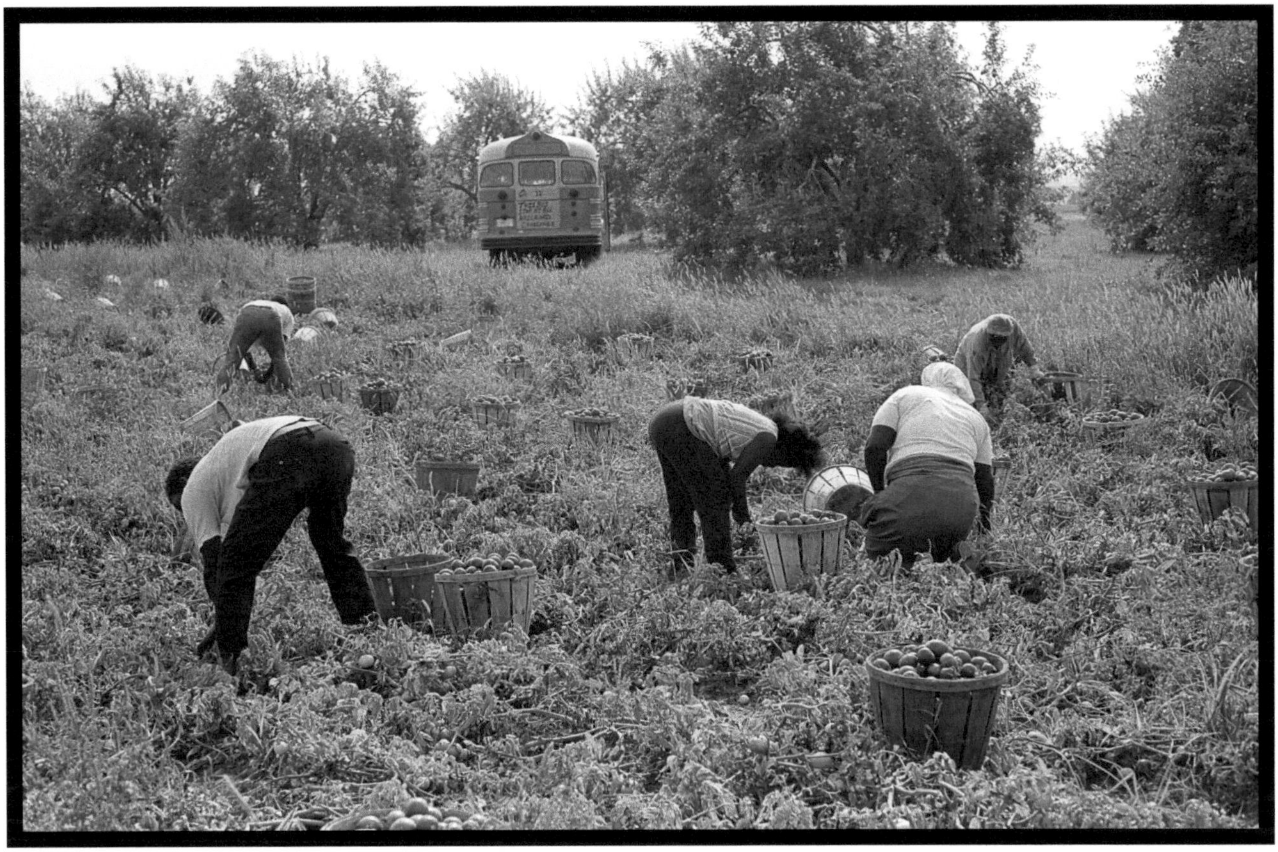

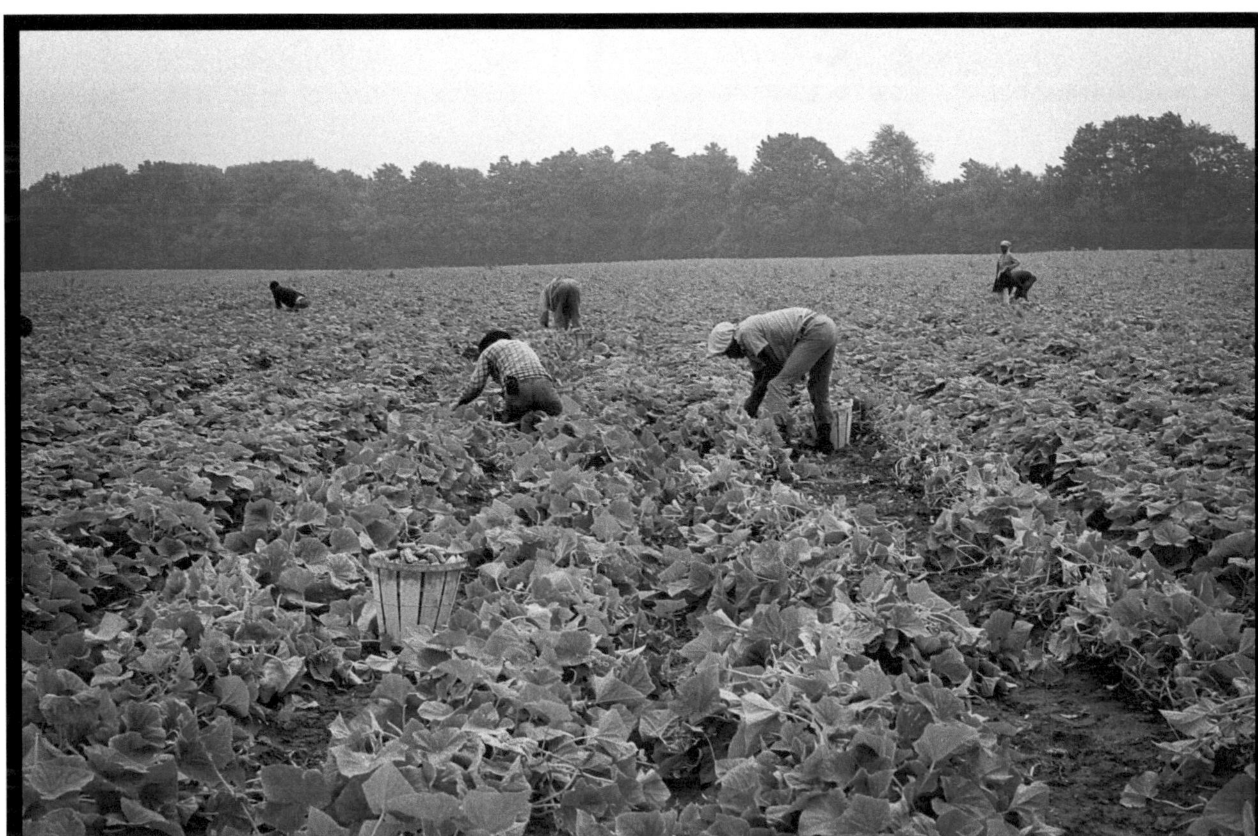

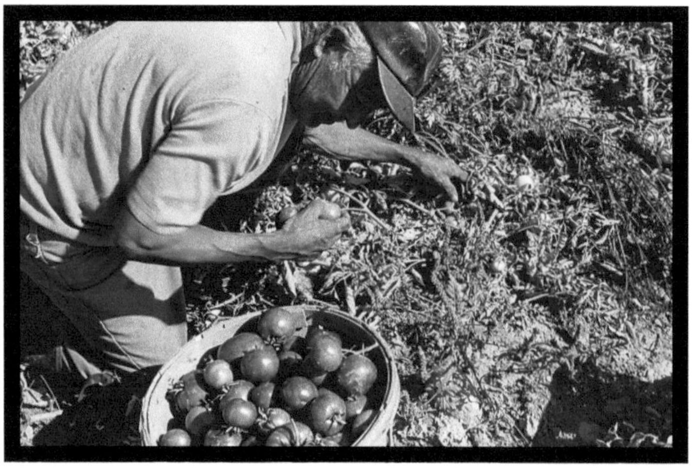
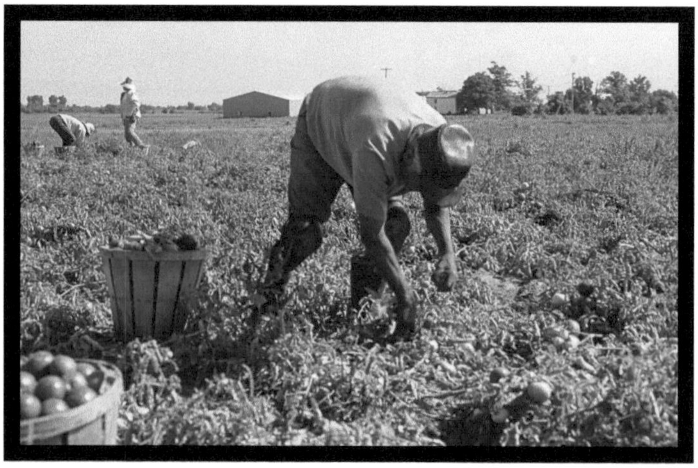
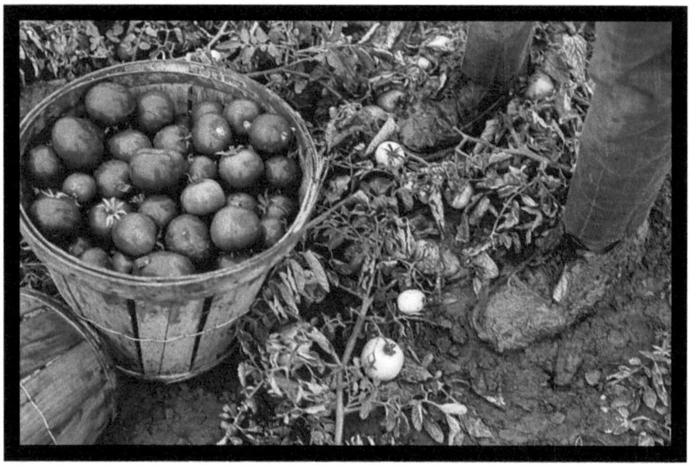
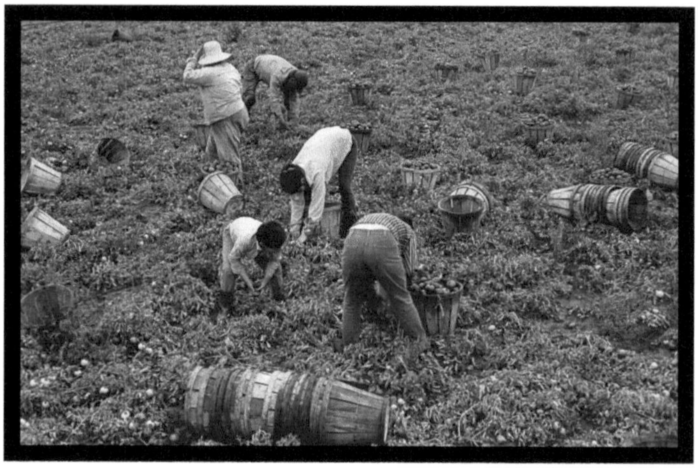
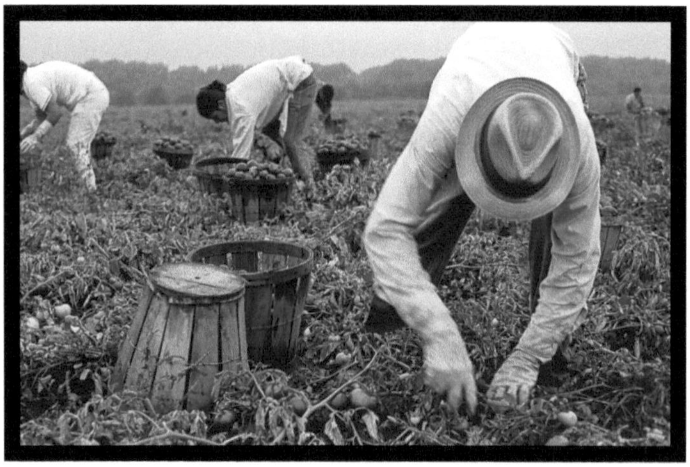
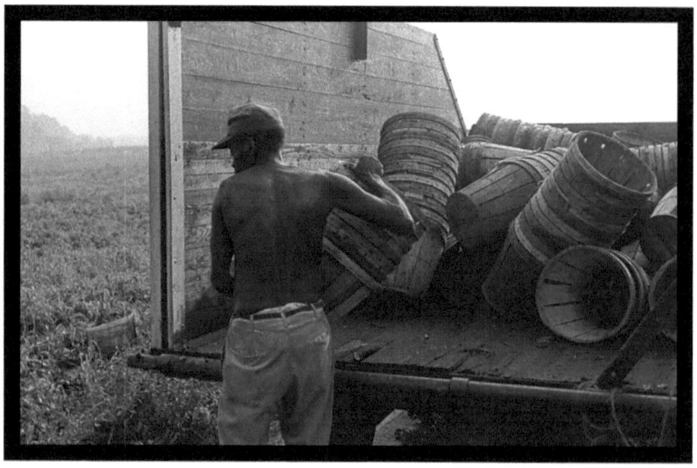

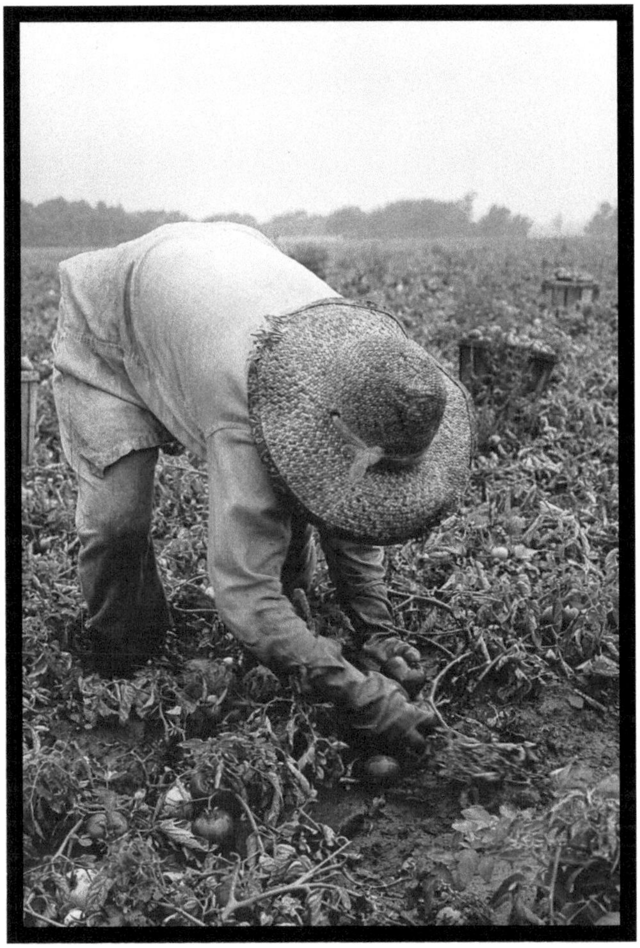

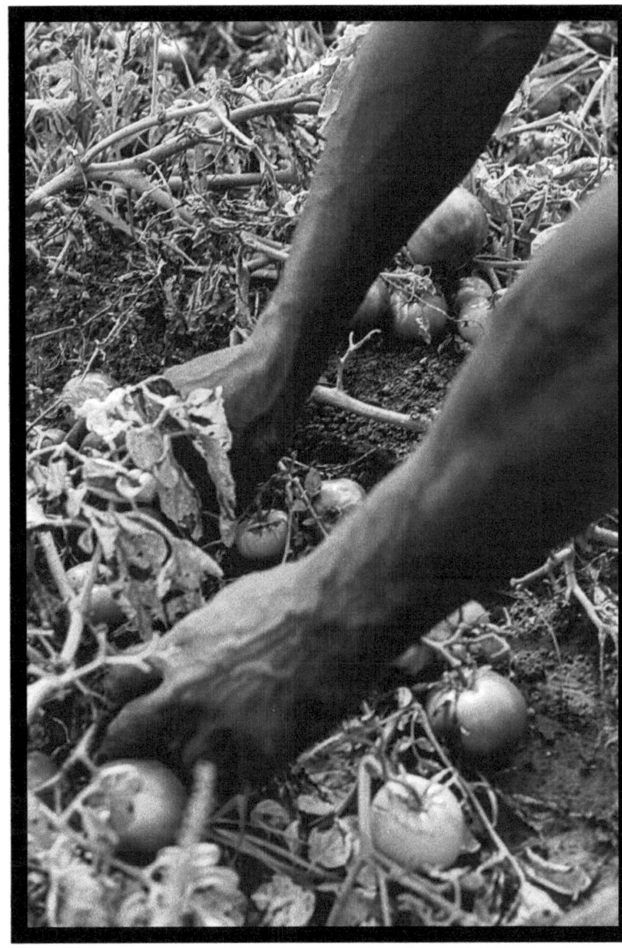

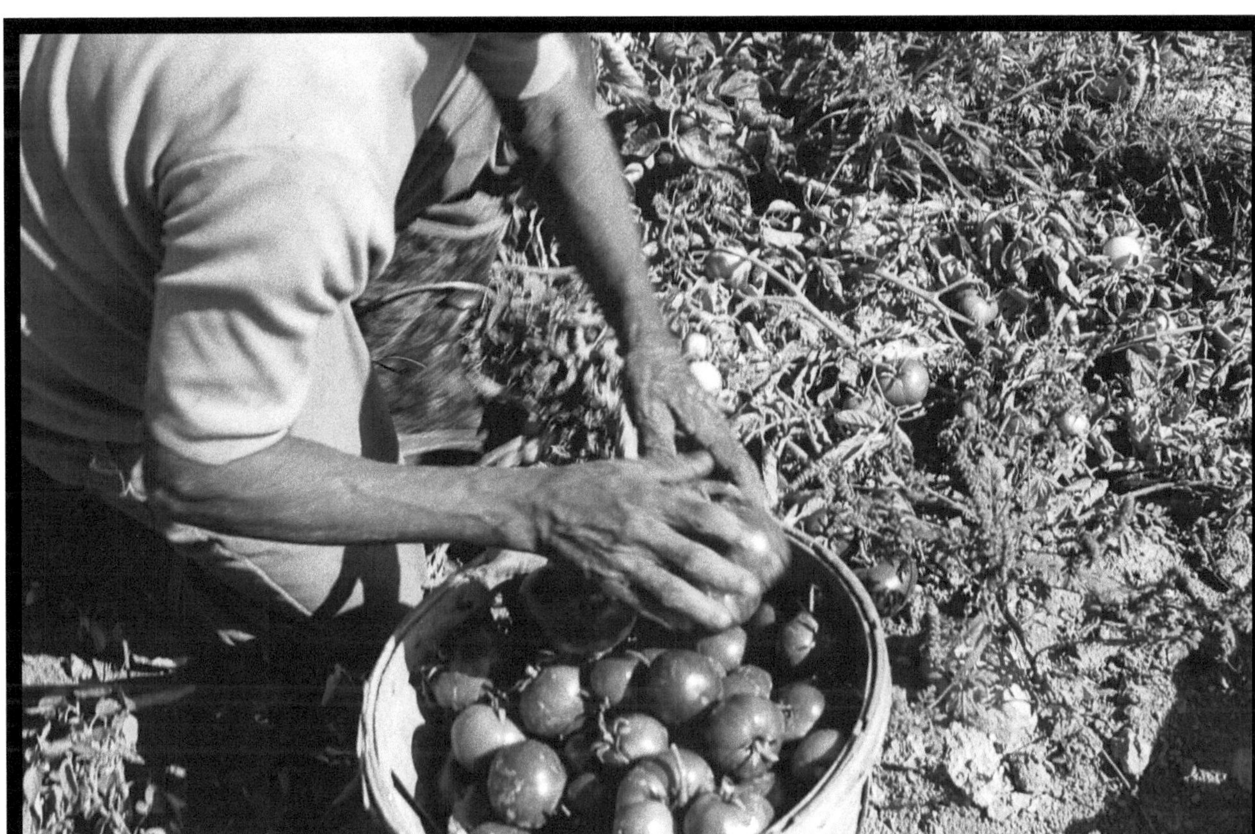

Tom Policano

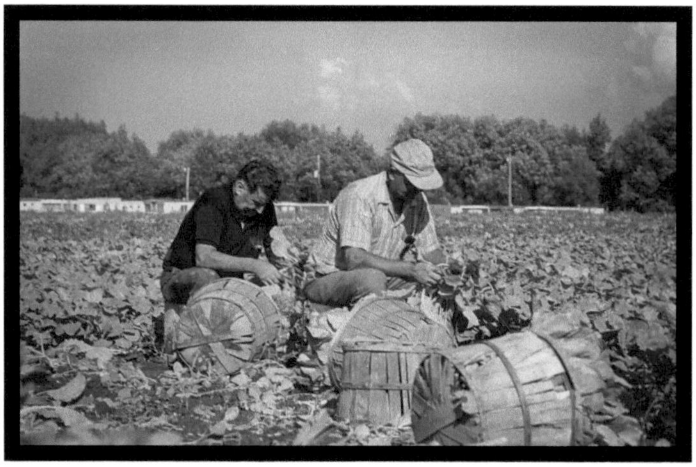

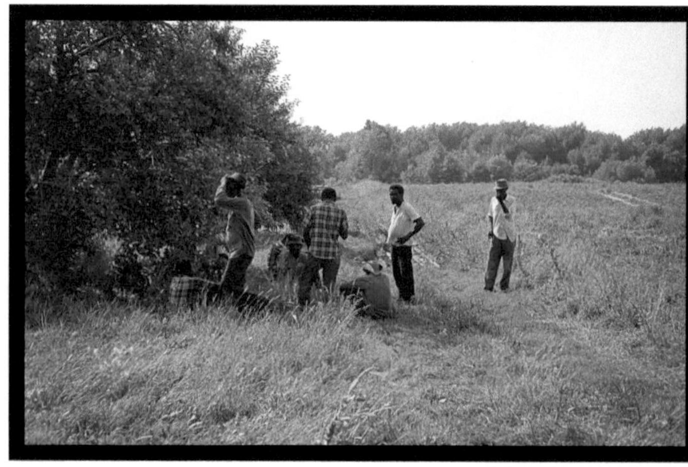

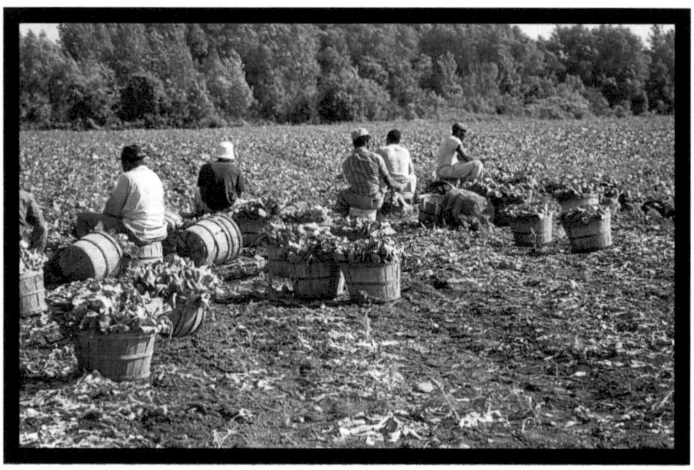

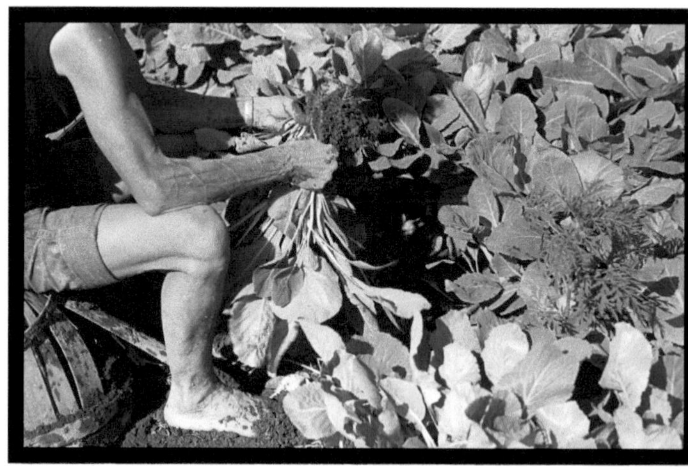

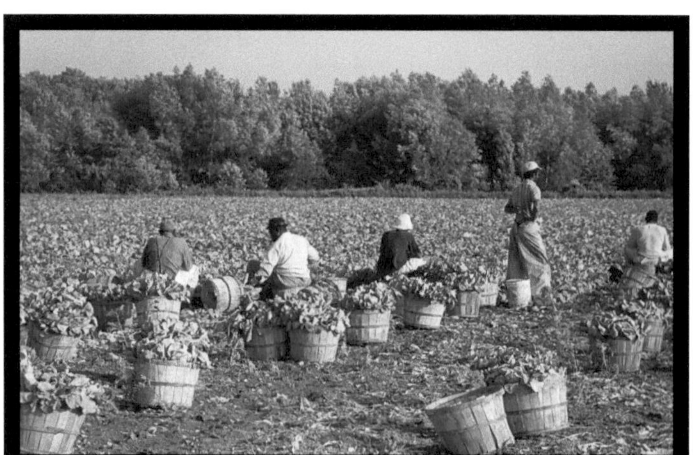

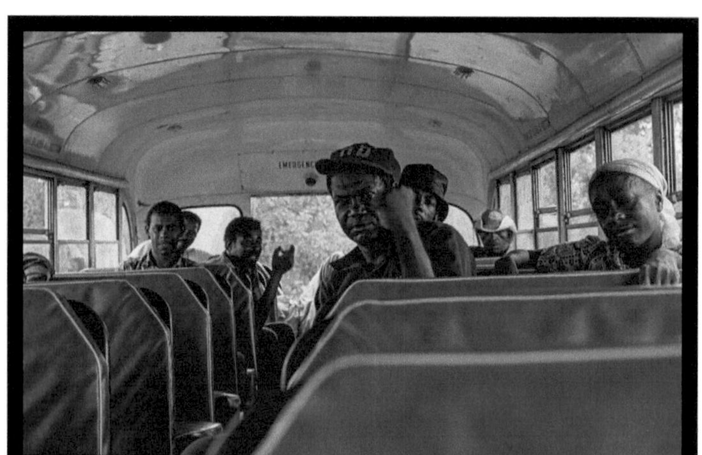

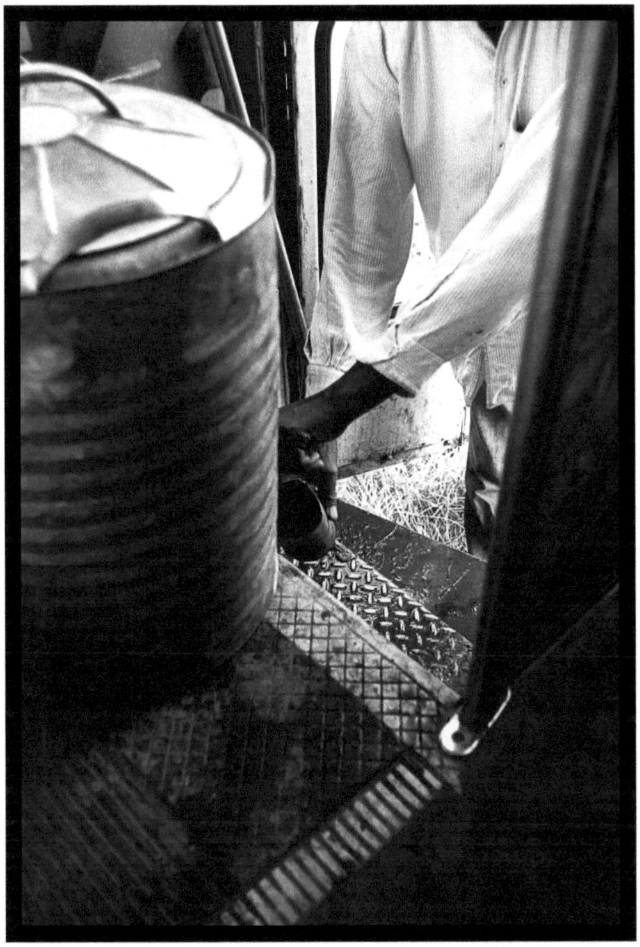

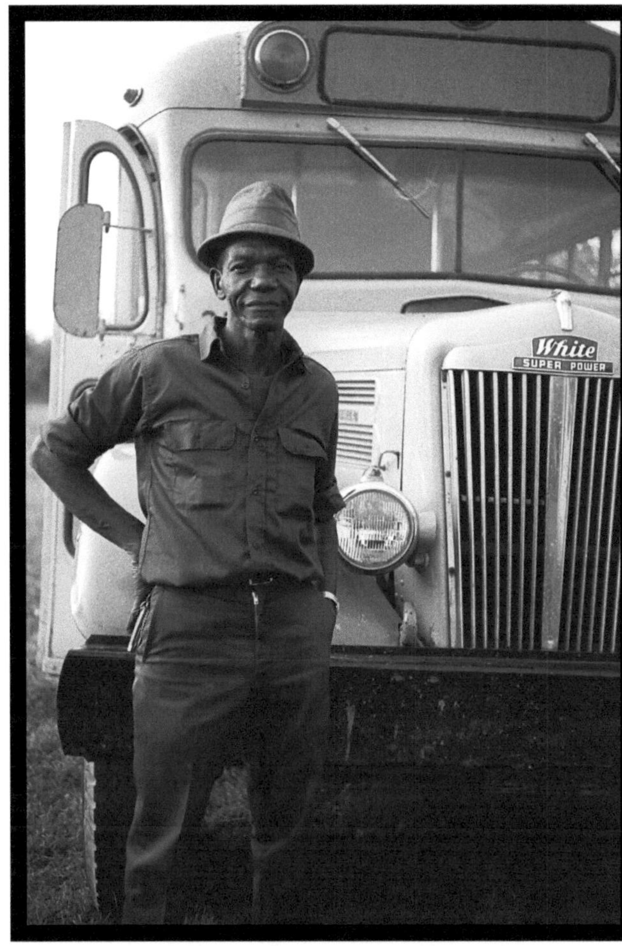

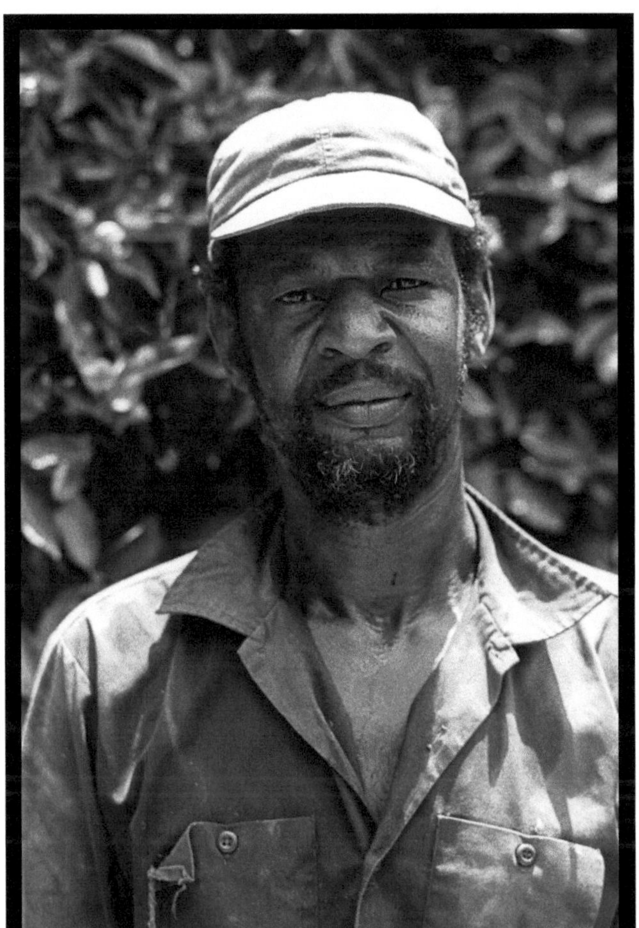

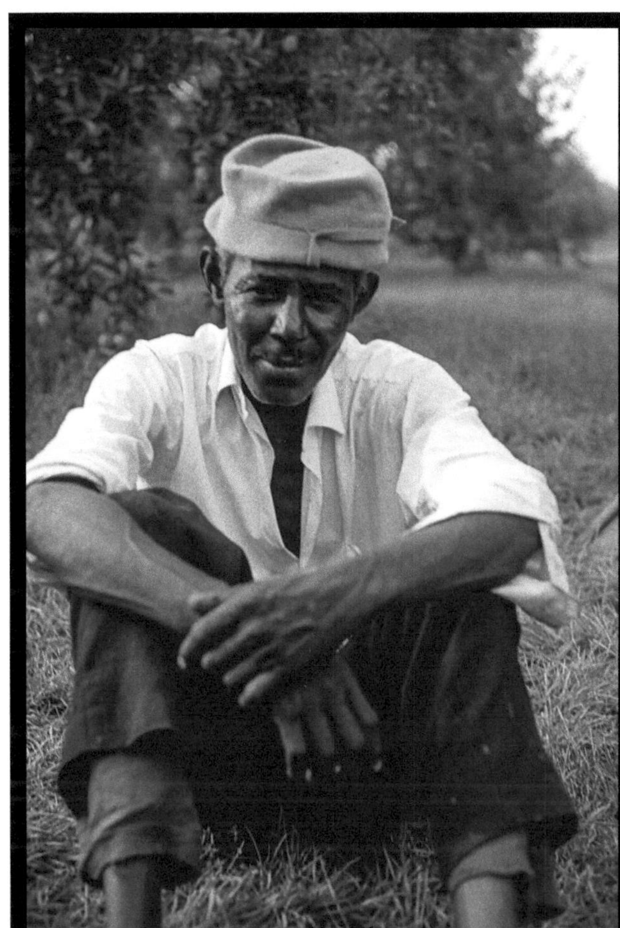

Tom Policano

North — In Camp

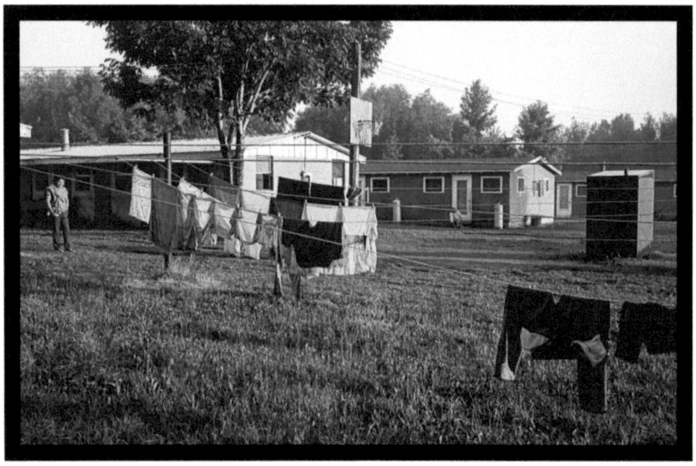

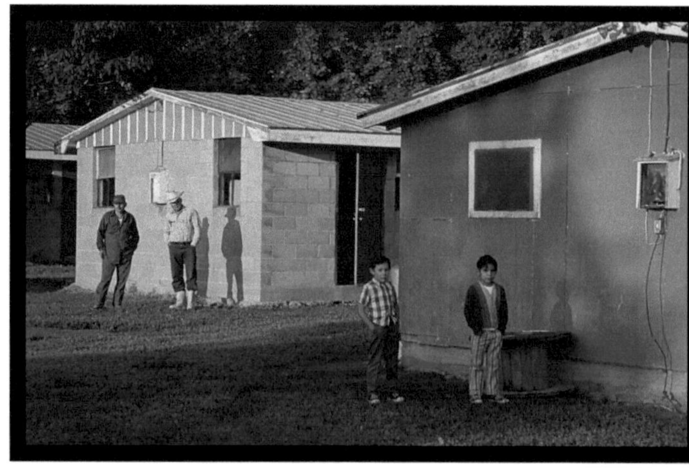

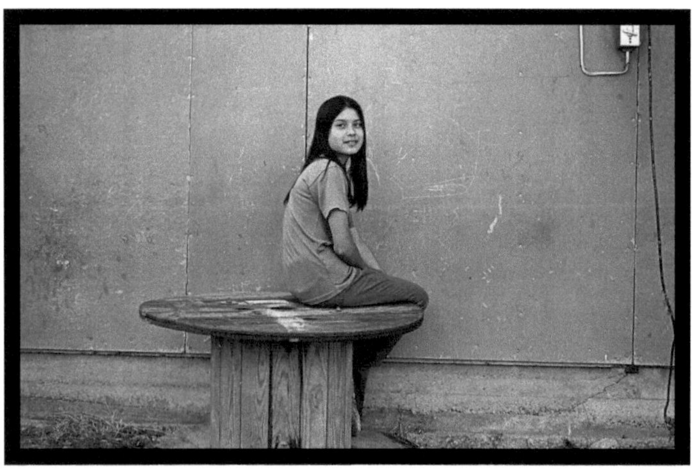

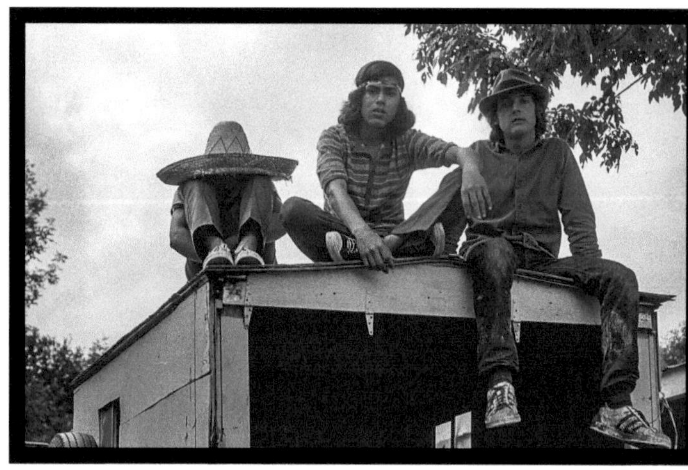

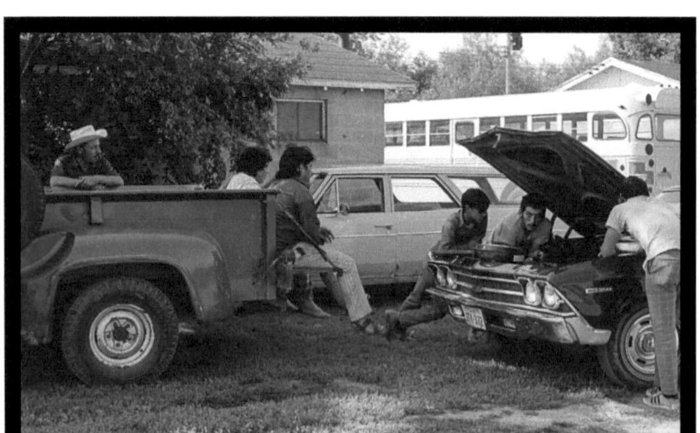

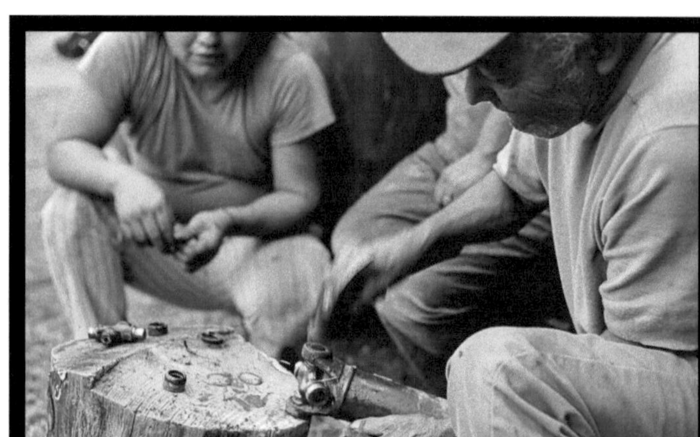

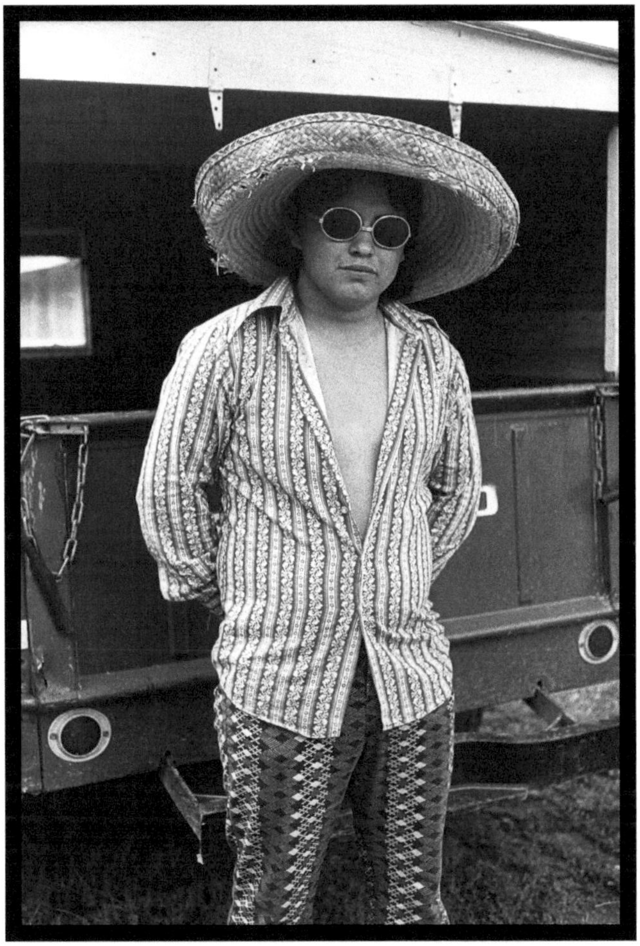

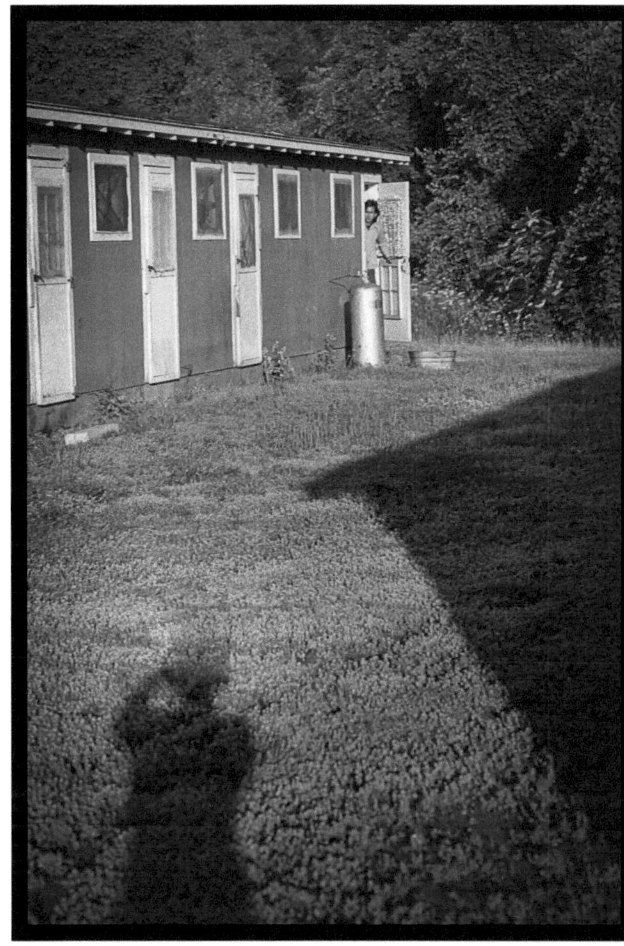

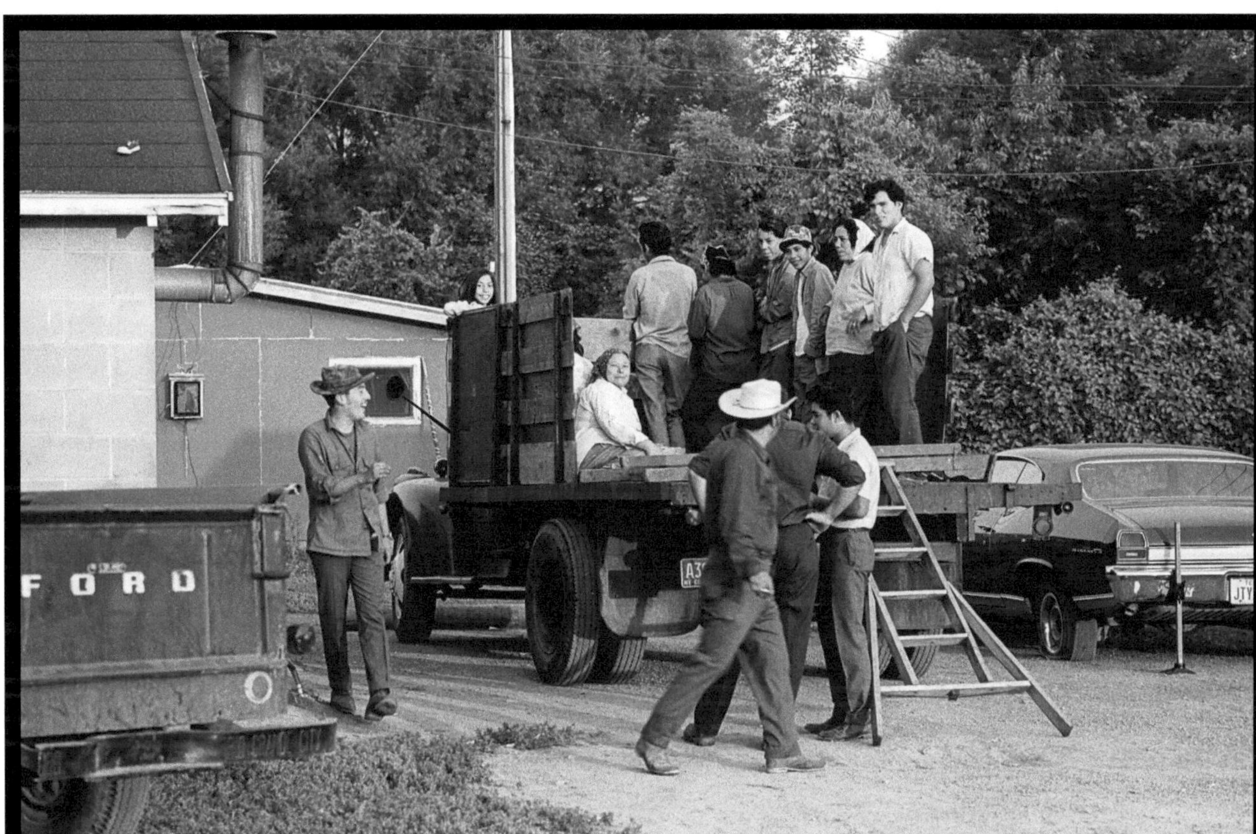

Tom Policano

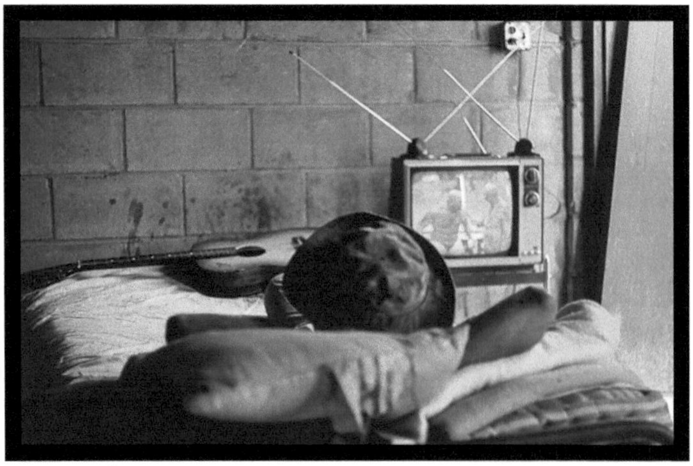

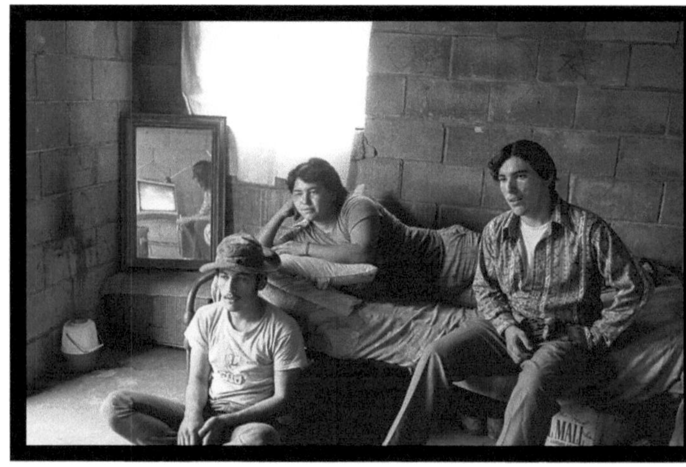

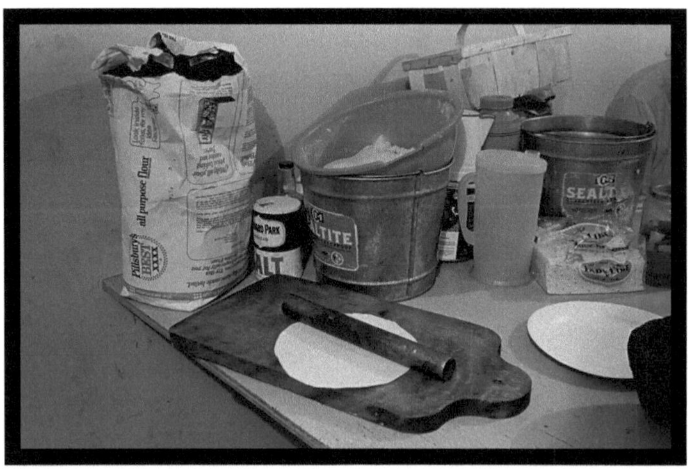
Pillsbury's BEST
all purpose flour
SEALTITE

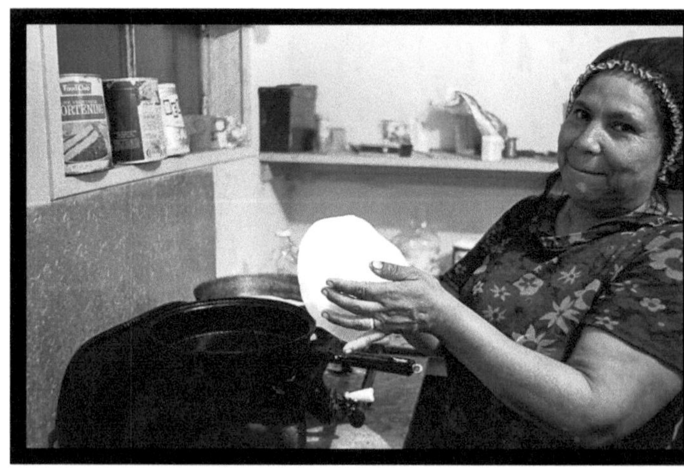

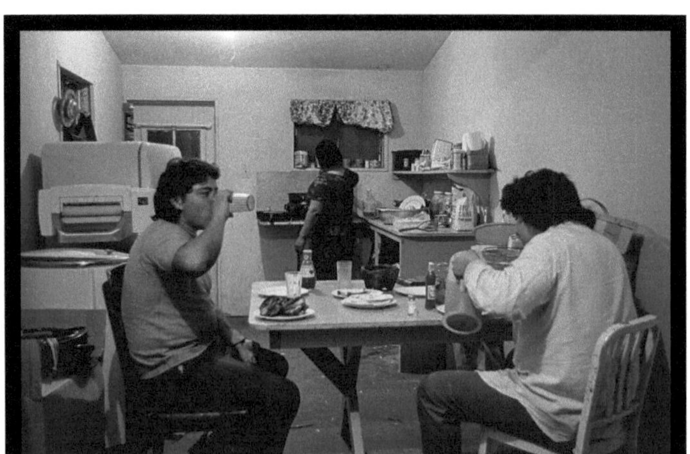

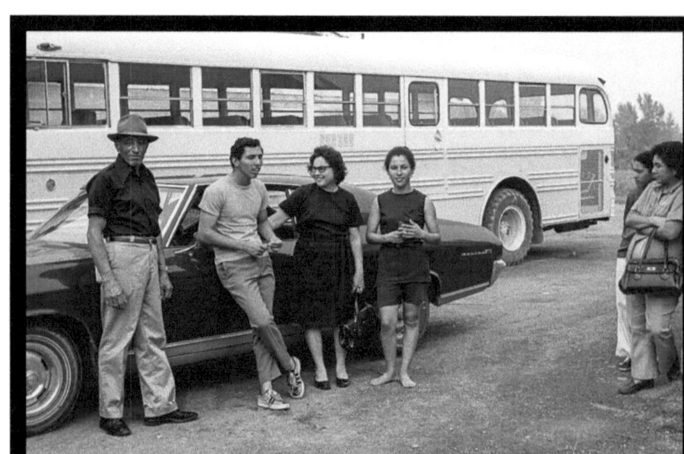

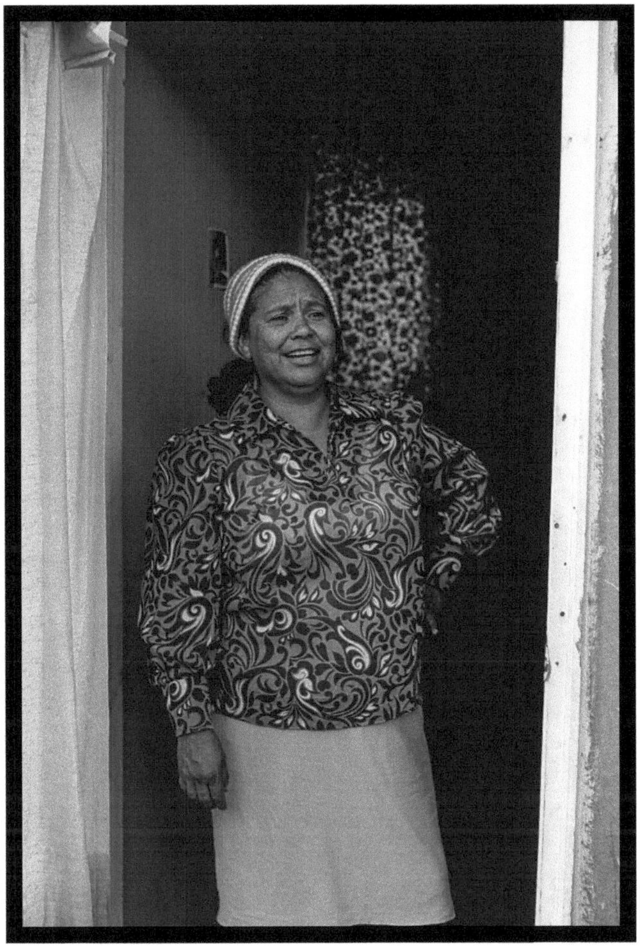

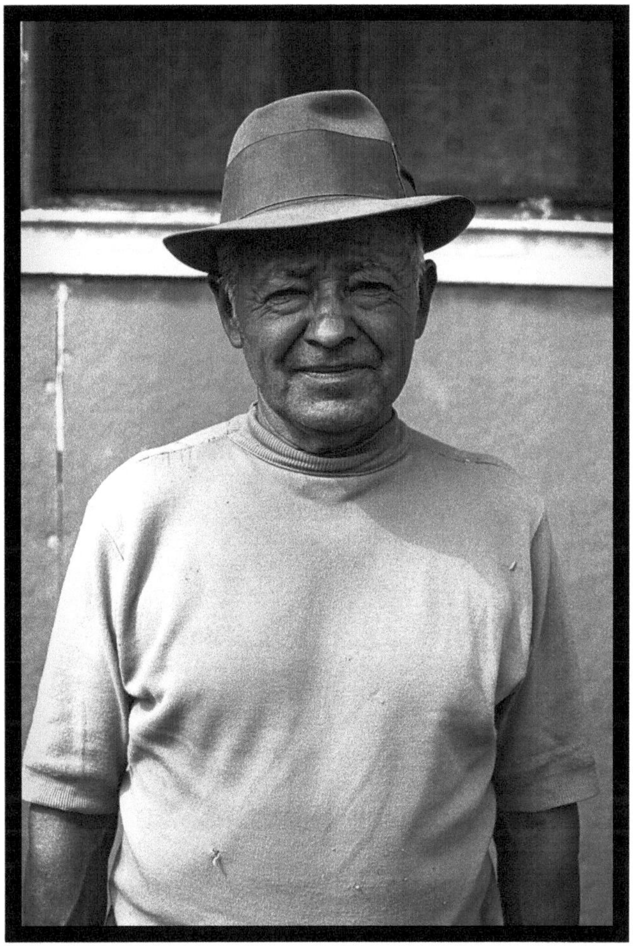

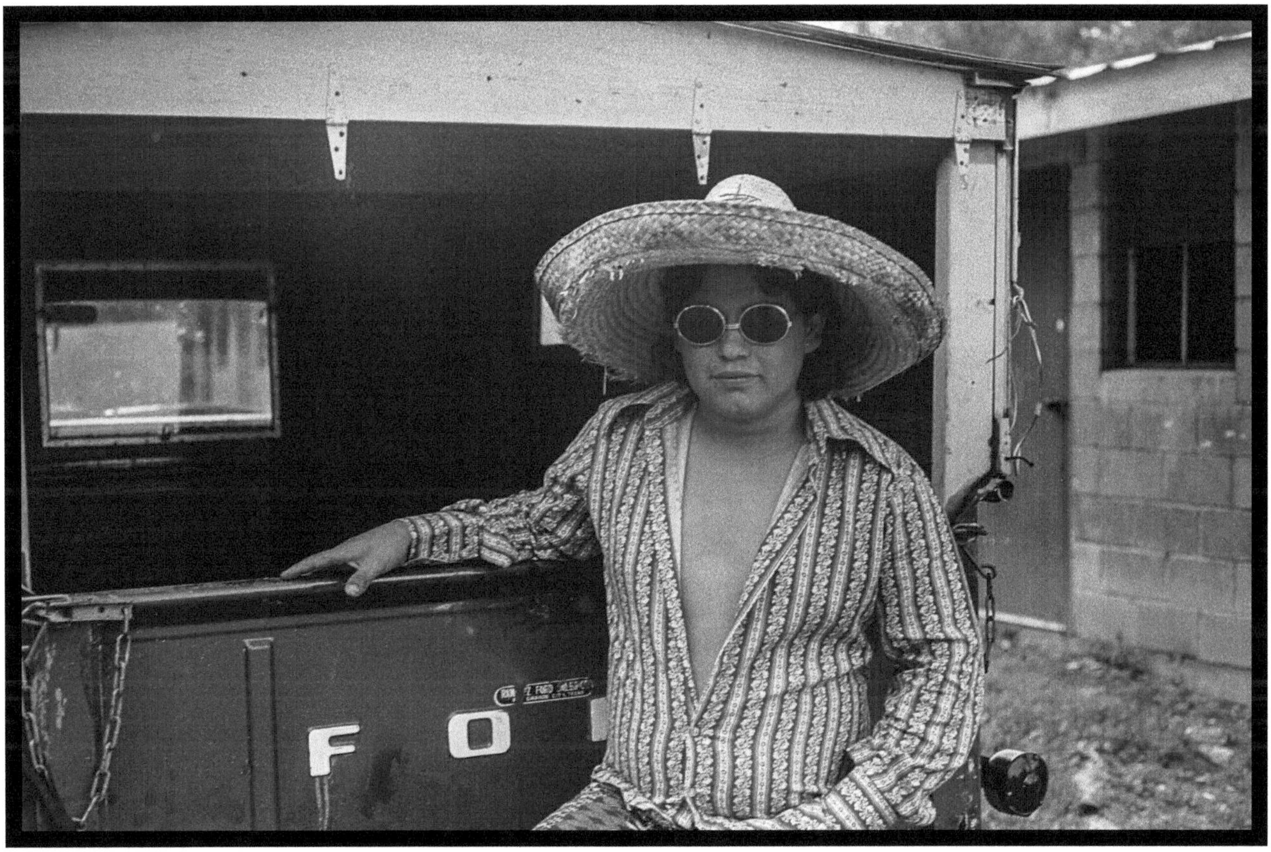

Tom Policano

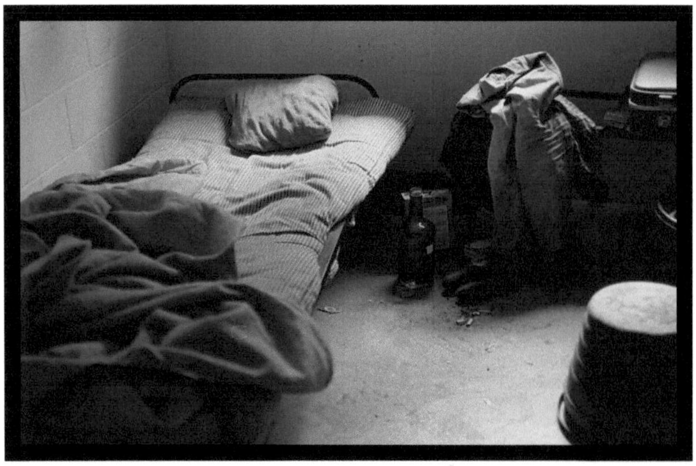

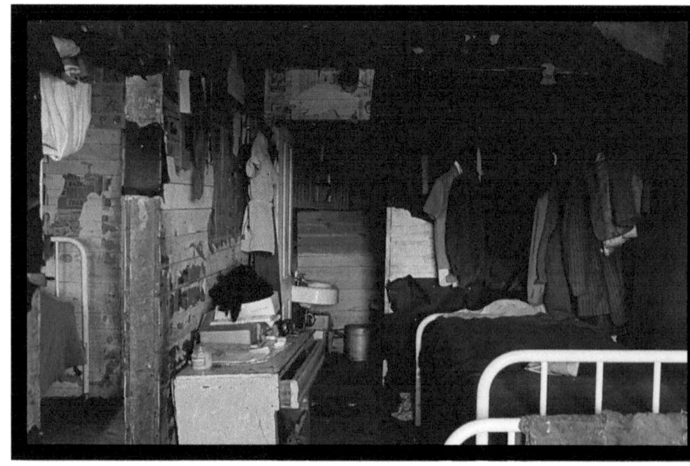

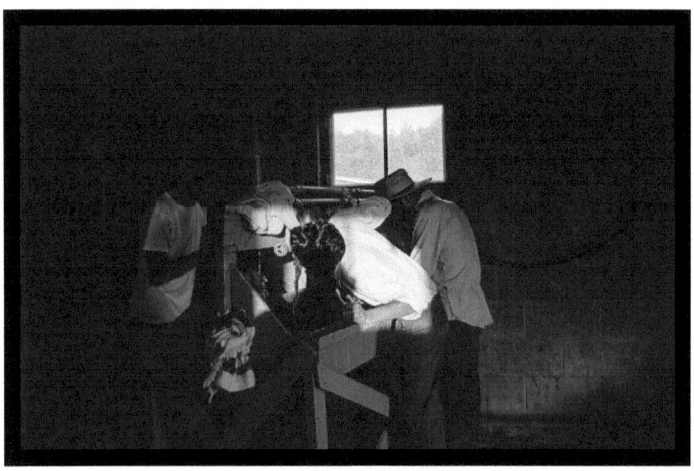

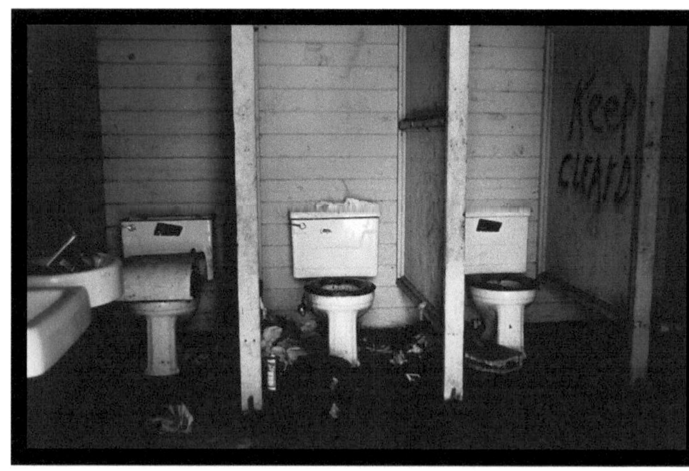
Keep

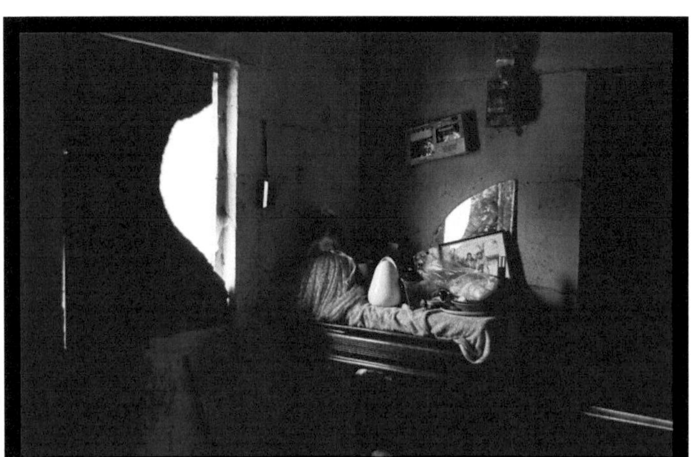

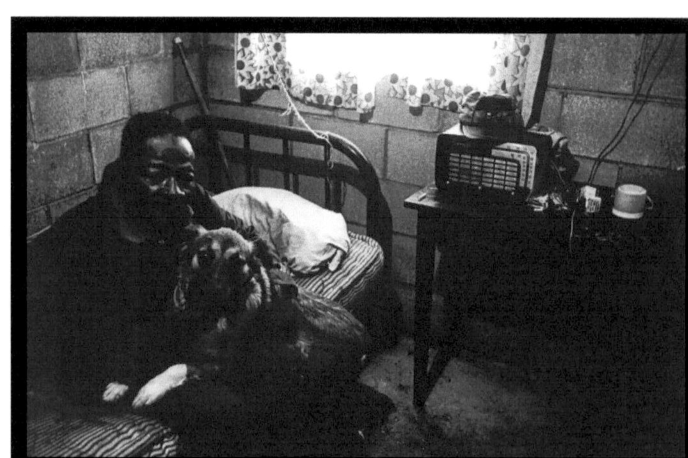

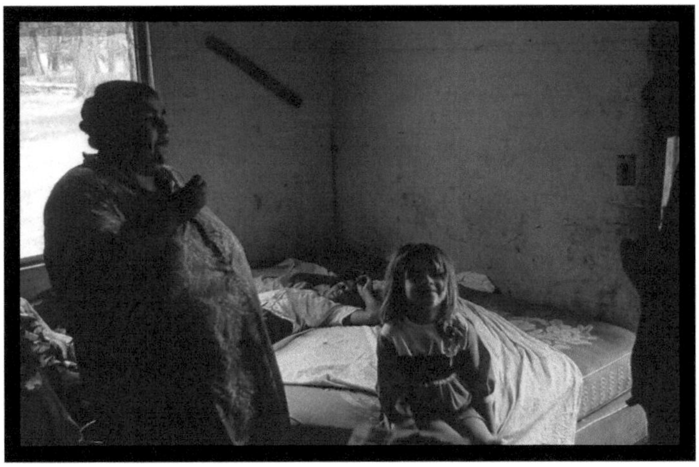

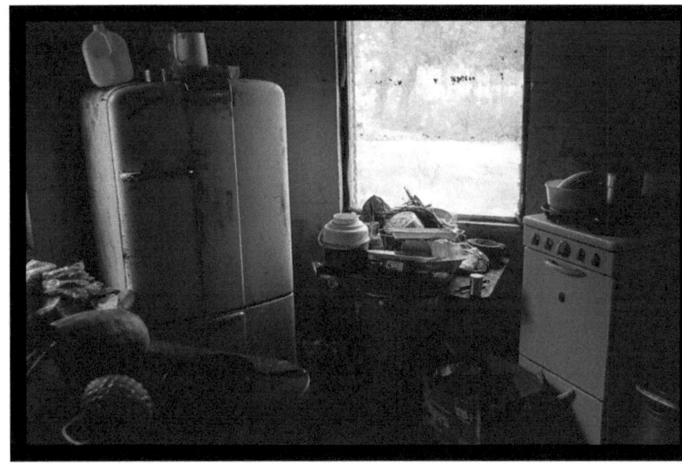

It got to rainin so bad one time we had to go out there in the car out there. That's the only dry place there was. We had to carry the stick. That was how bad it was now. Ana they got about close to about twelve rats in there really. Twelve!

That's includin the little baby ones, too,

No, you wrong, man. There's some rats down there. THAT LONG. Down there, what was walkin' on my bed. Man, an they big ones too! They look like squirrels.

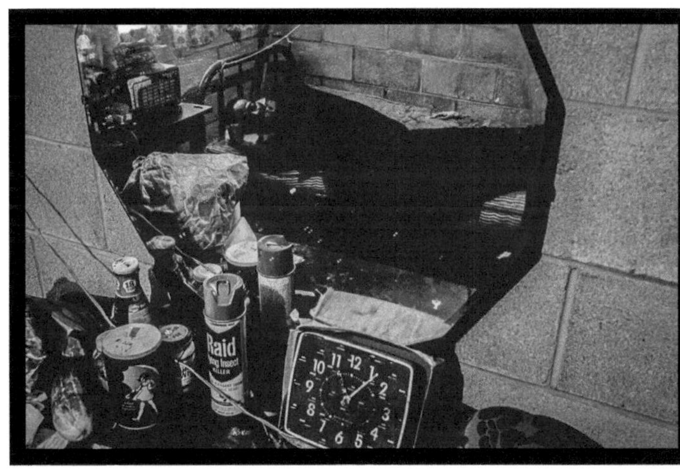

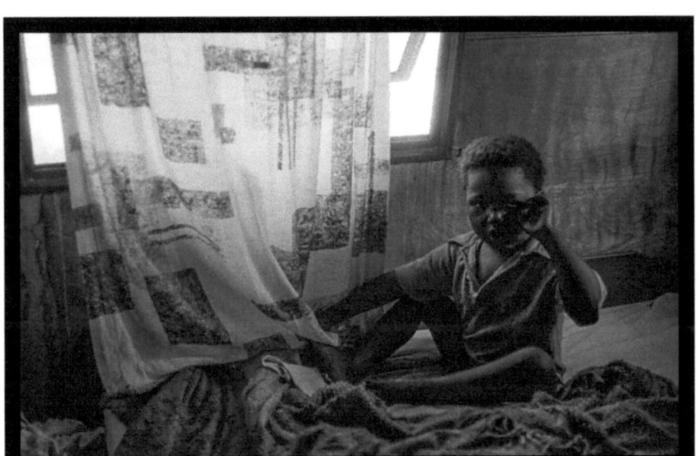

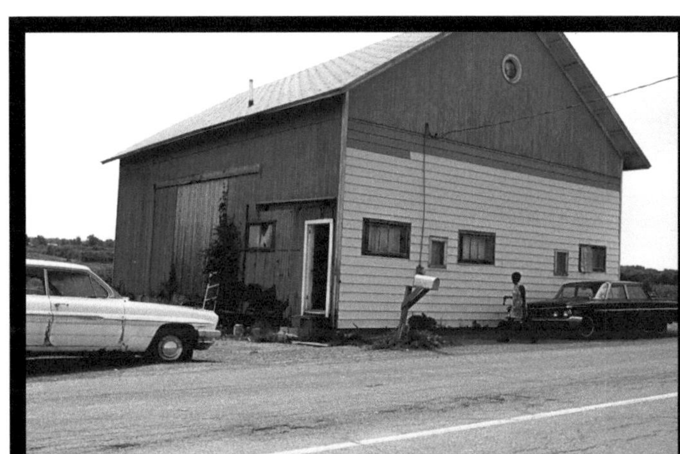

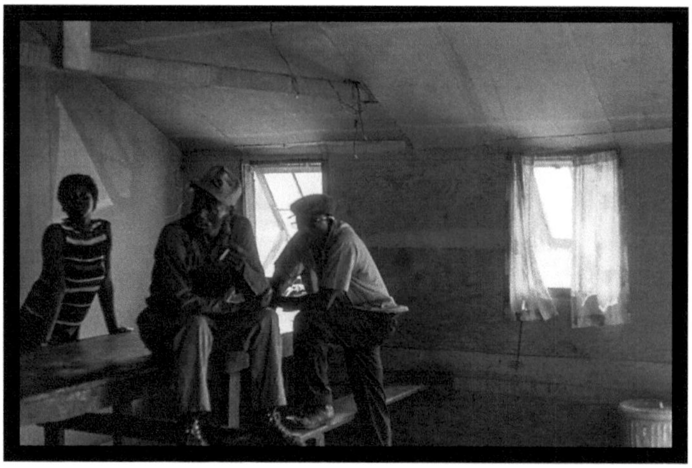

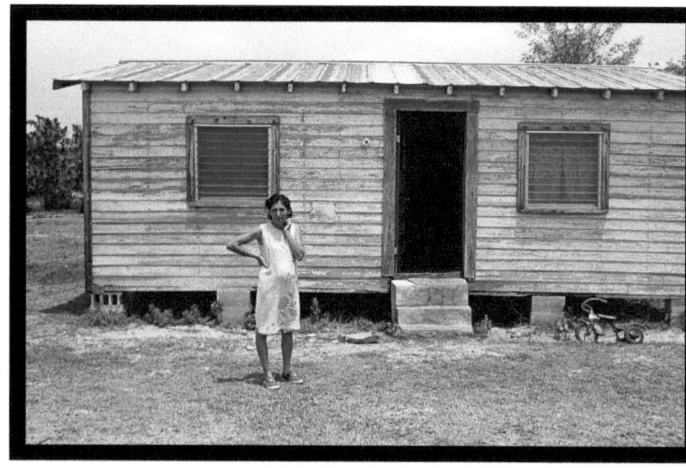

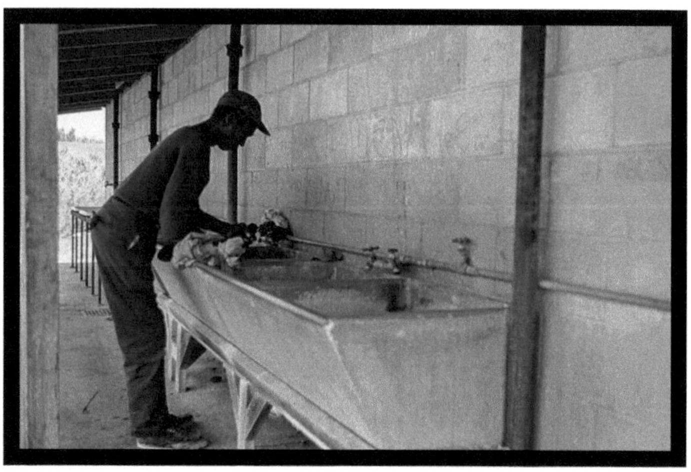

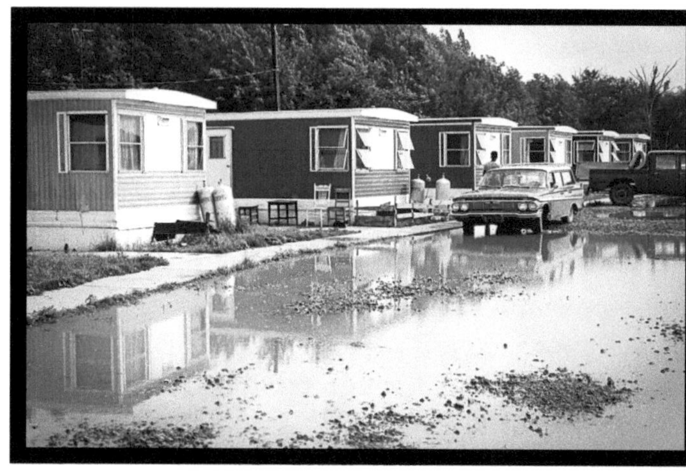

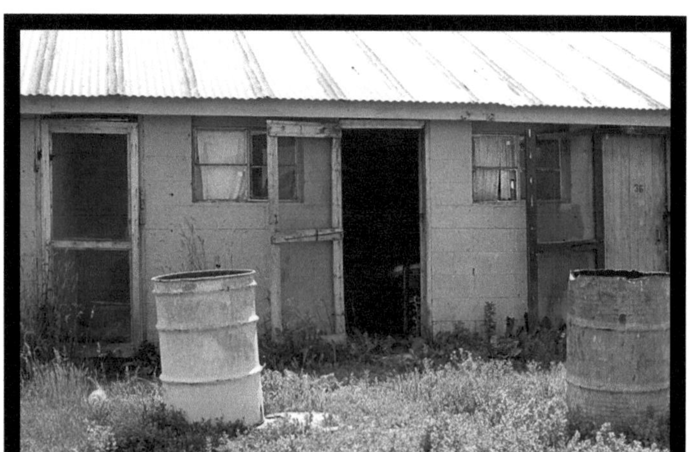

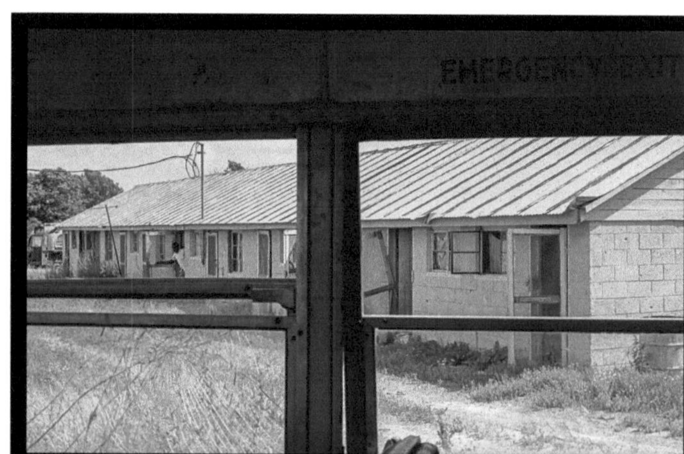
EMERGENCY

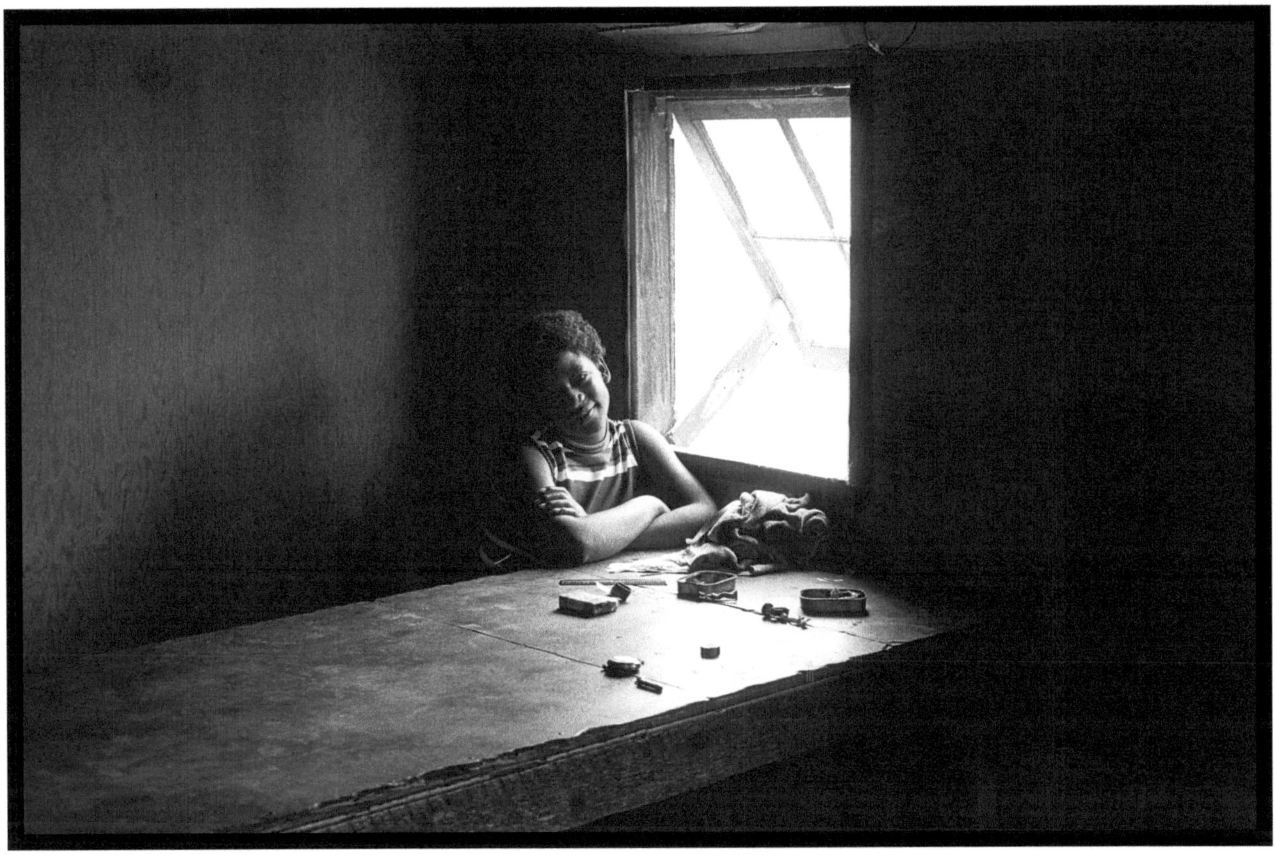

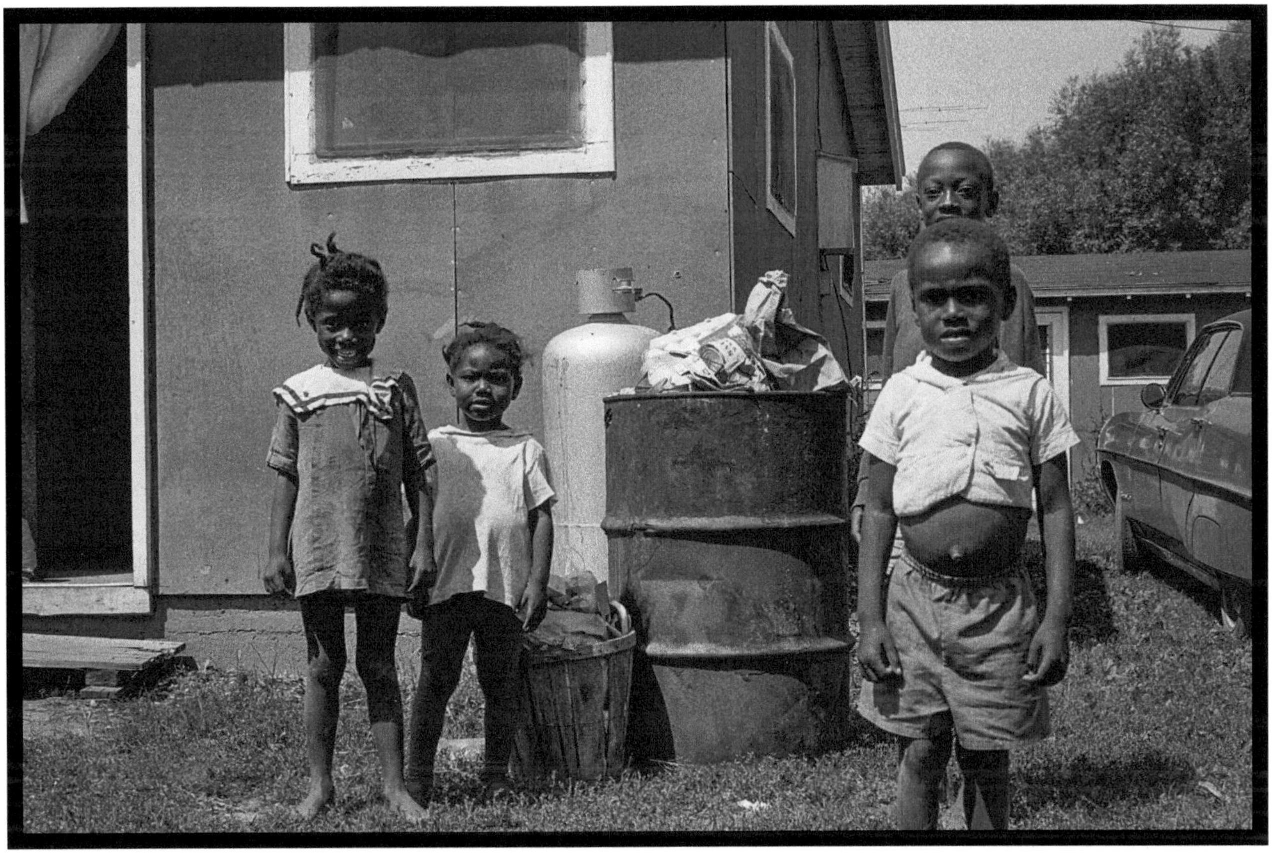

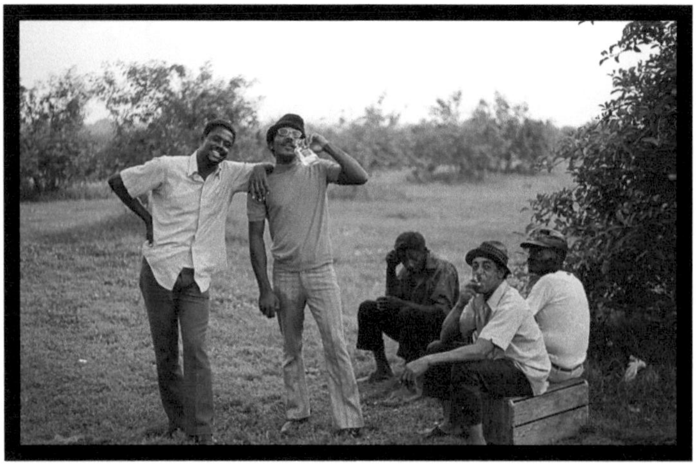
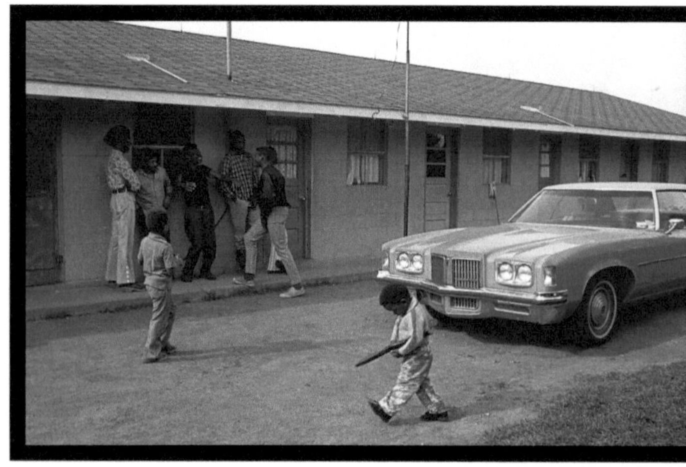
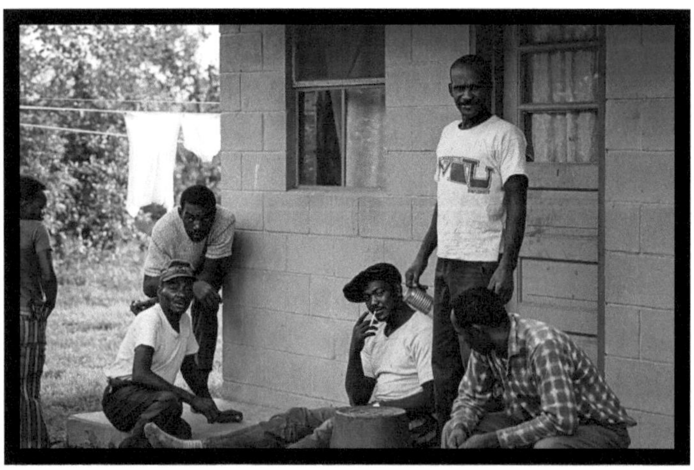
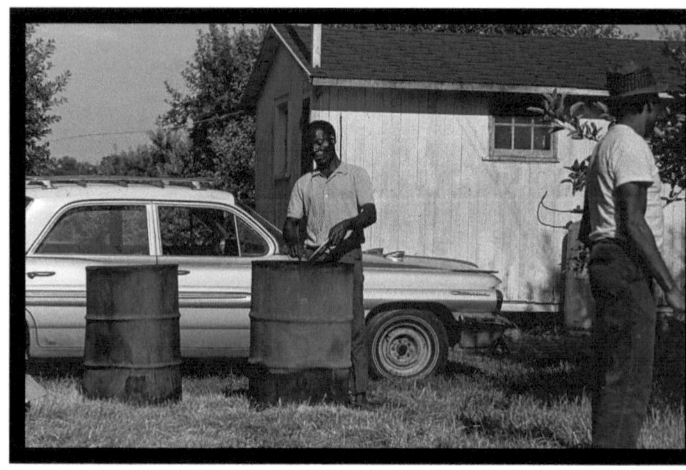
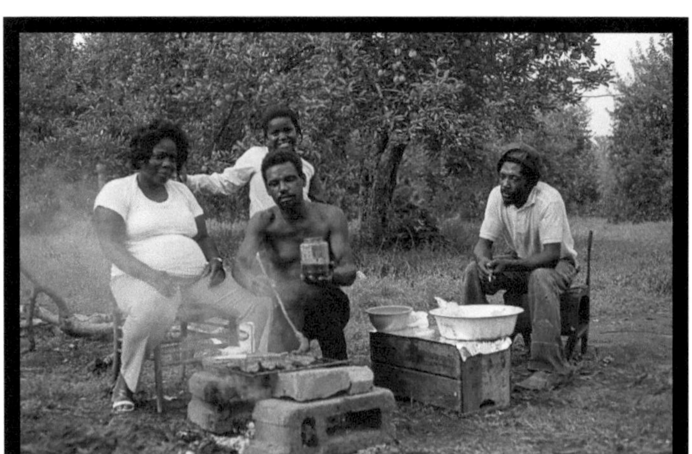
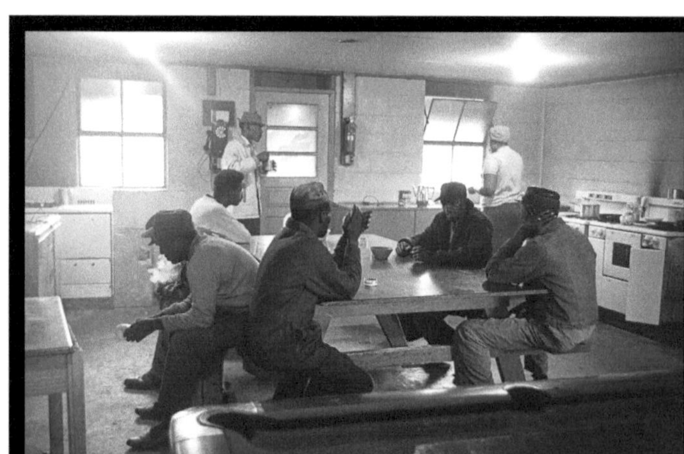

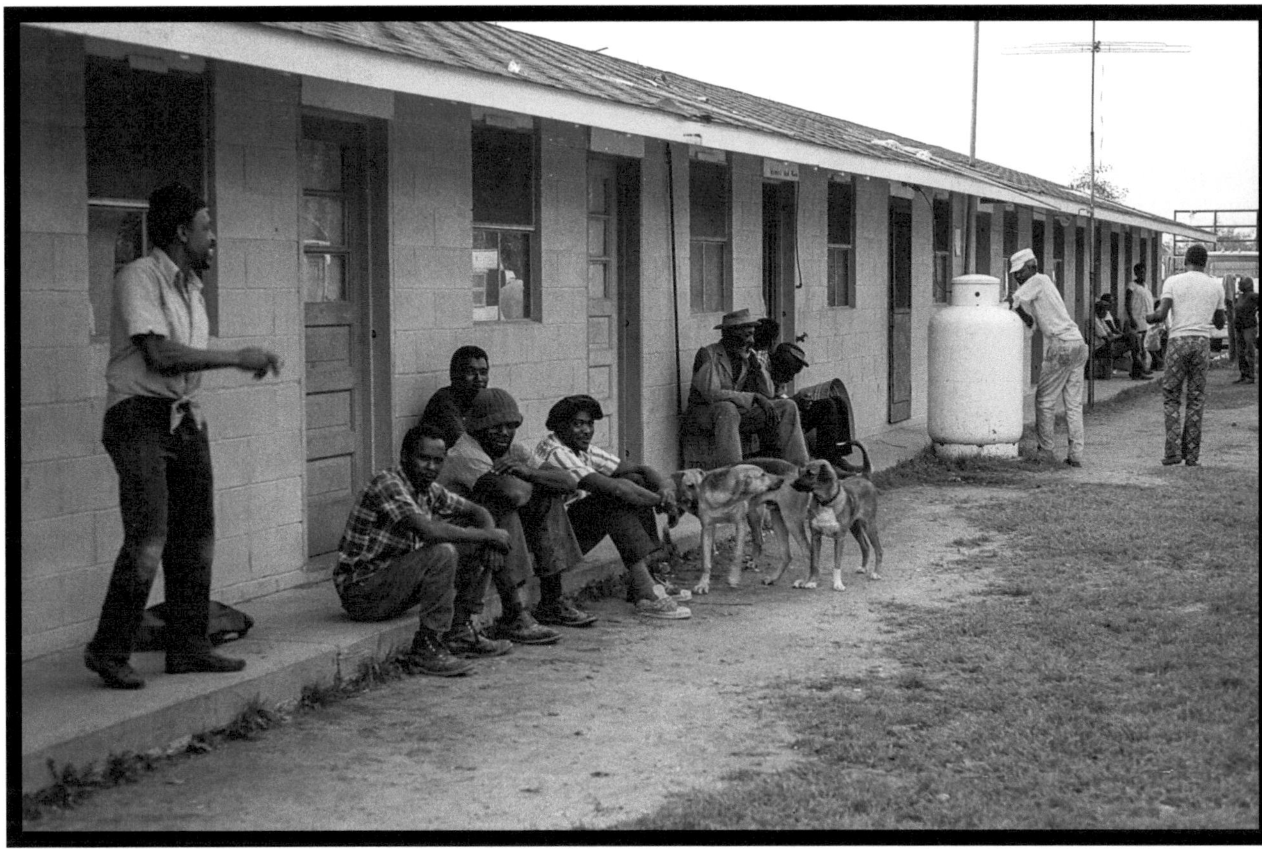

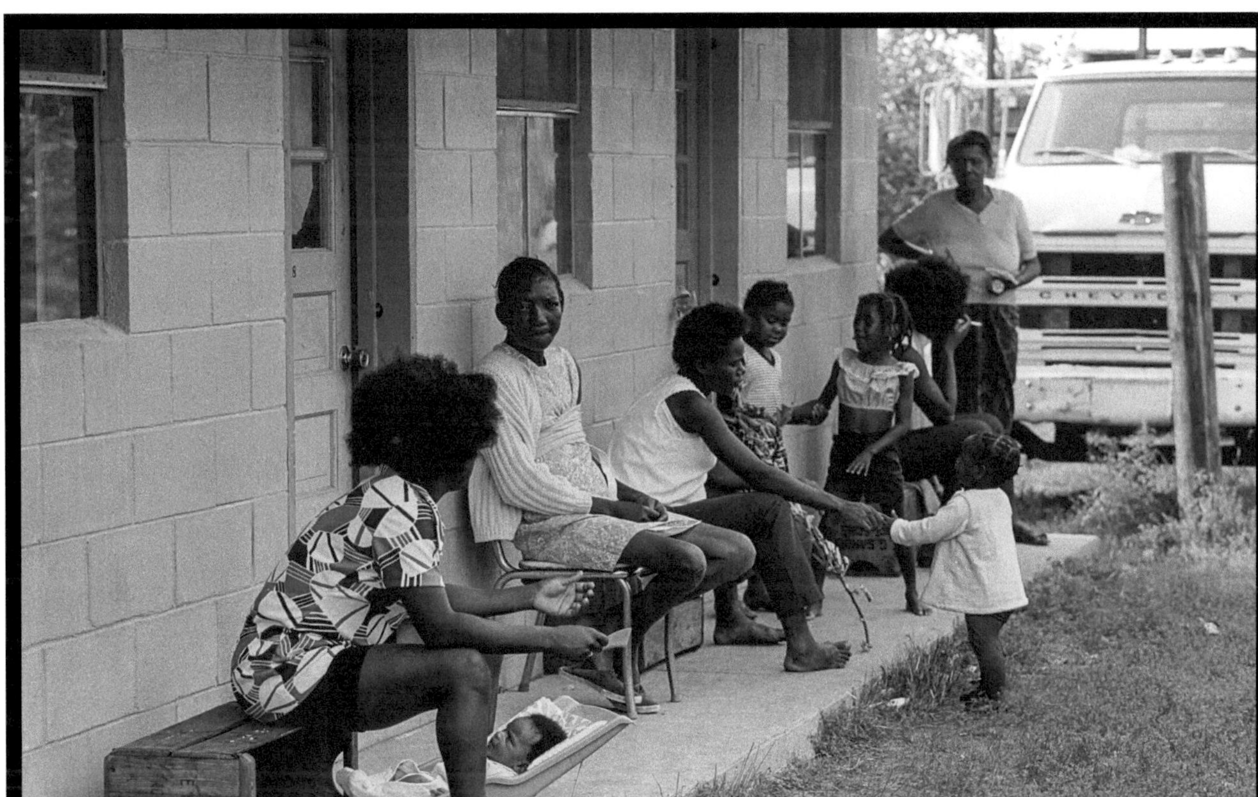

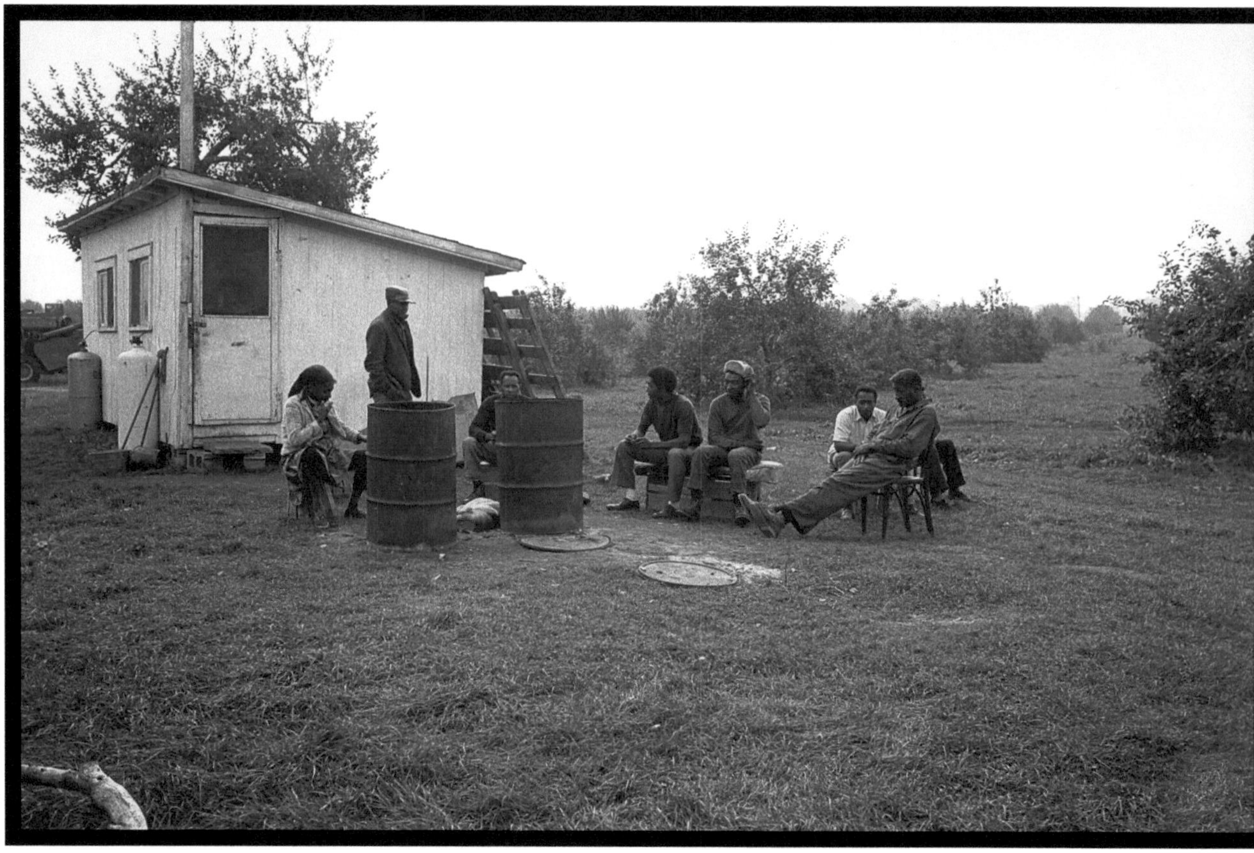

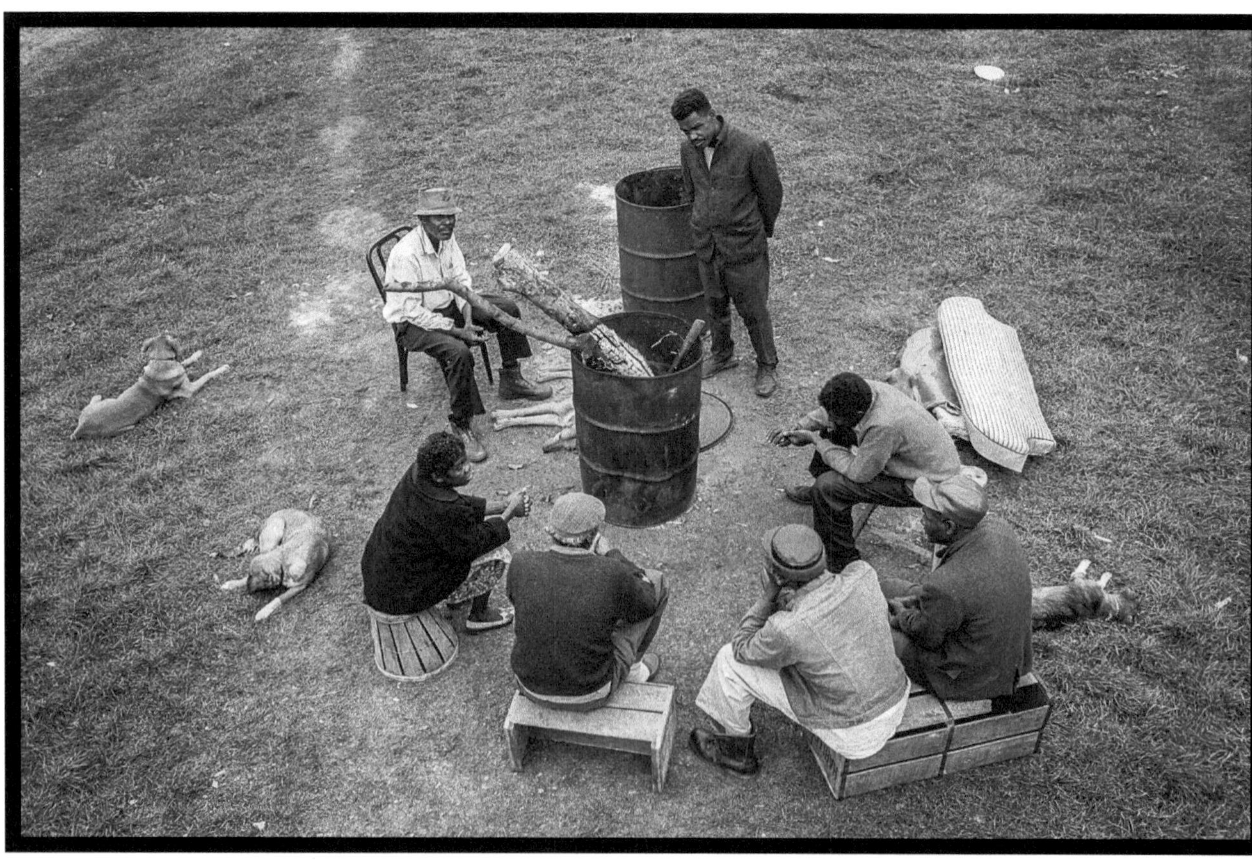

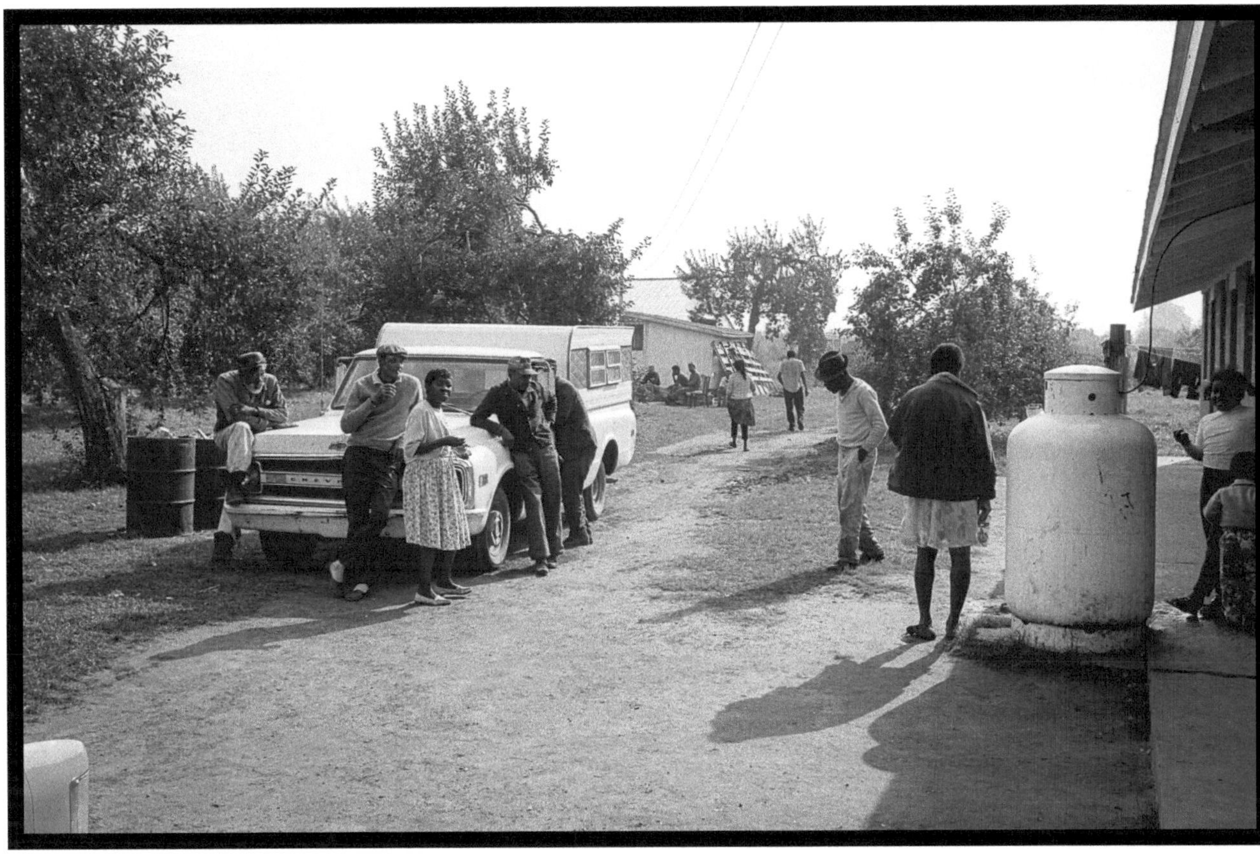

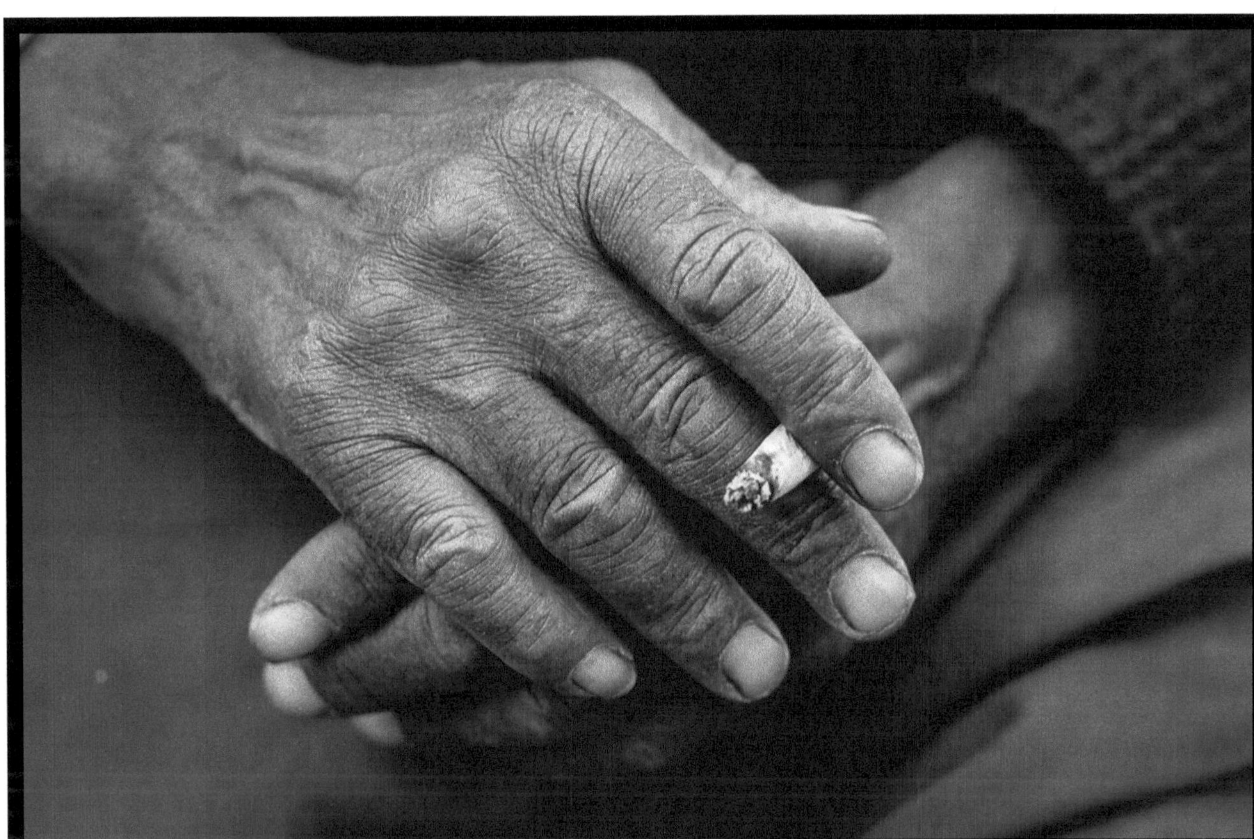

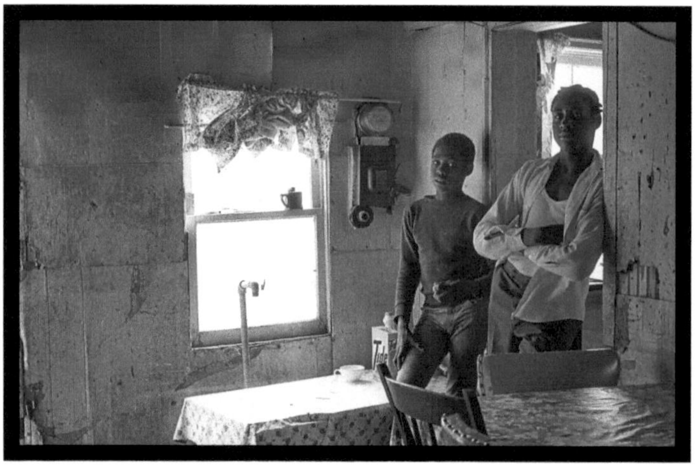

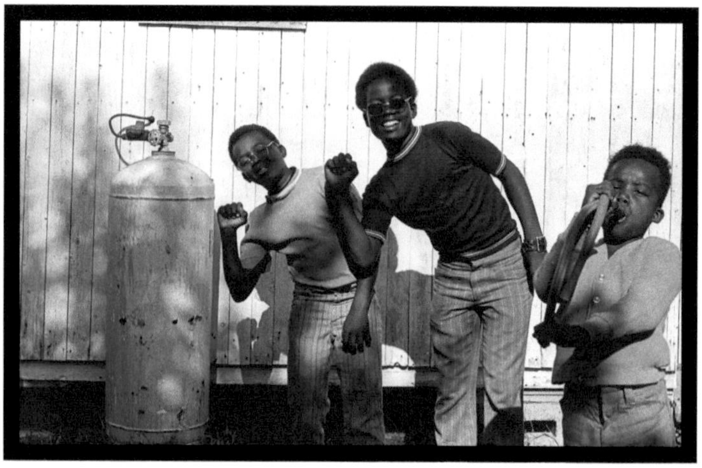

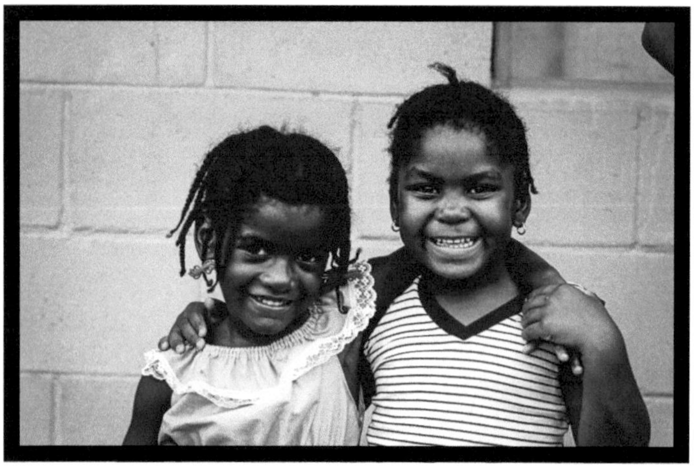

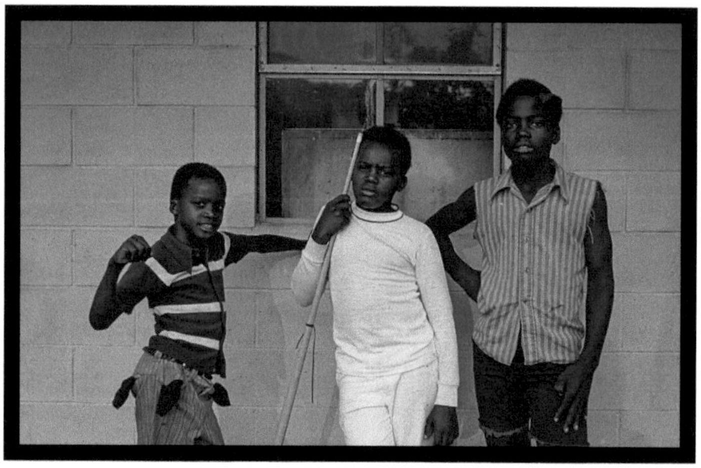

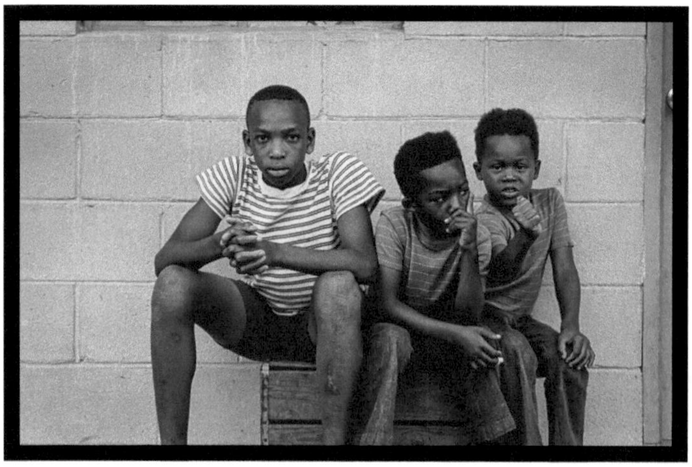

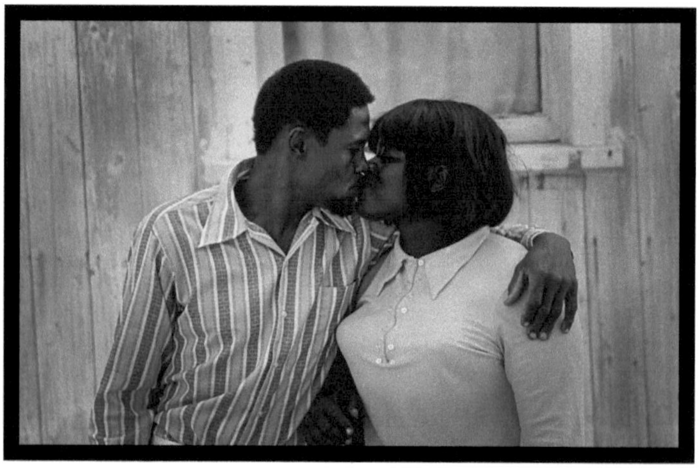

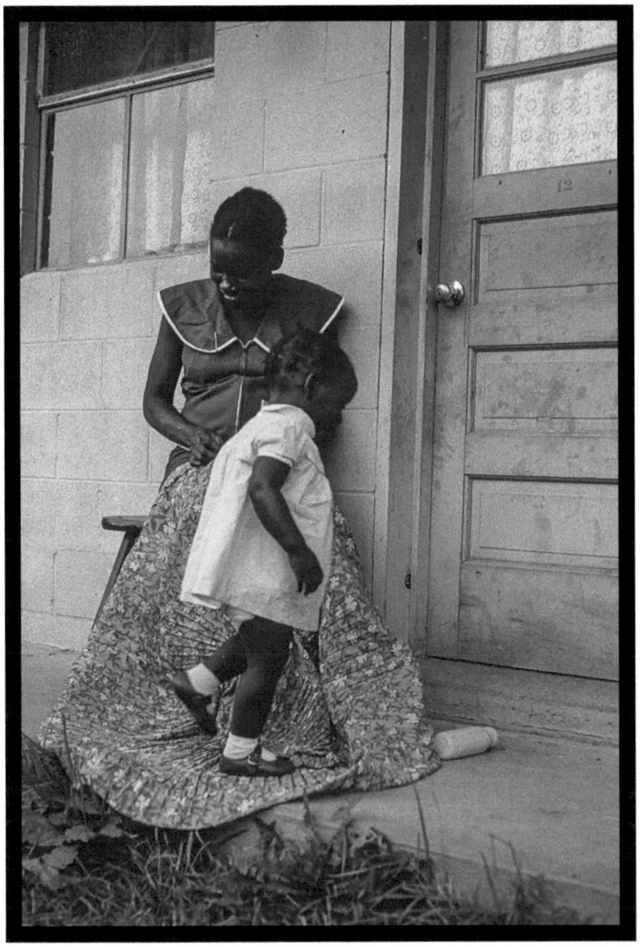

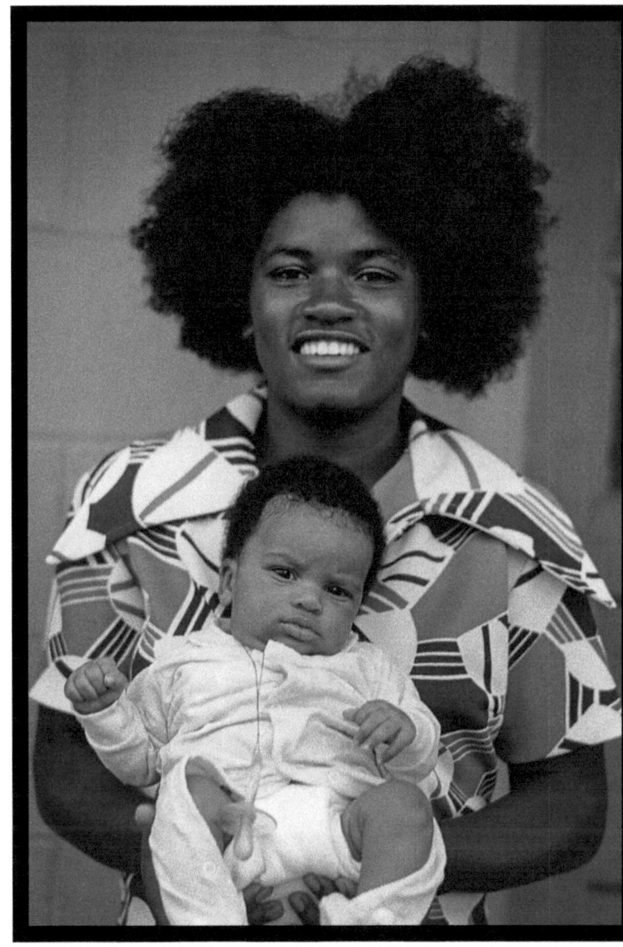

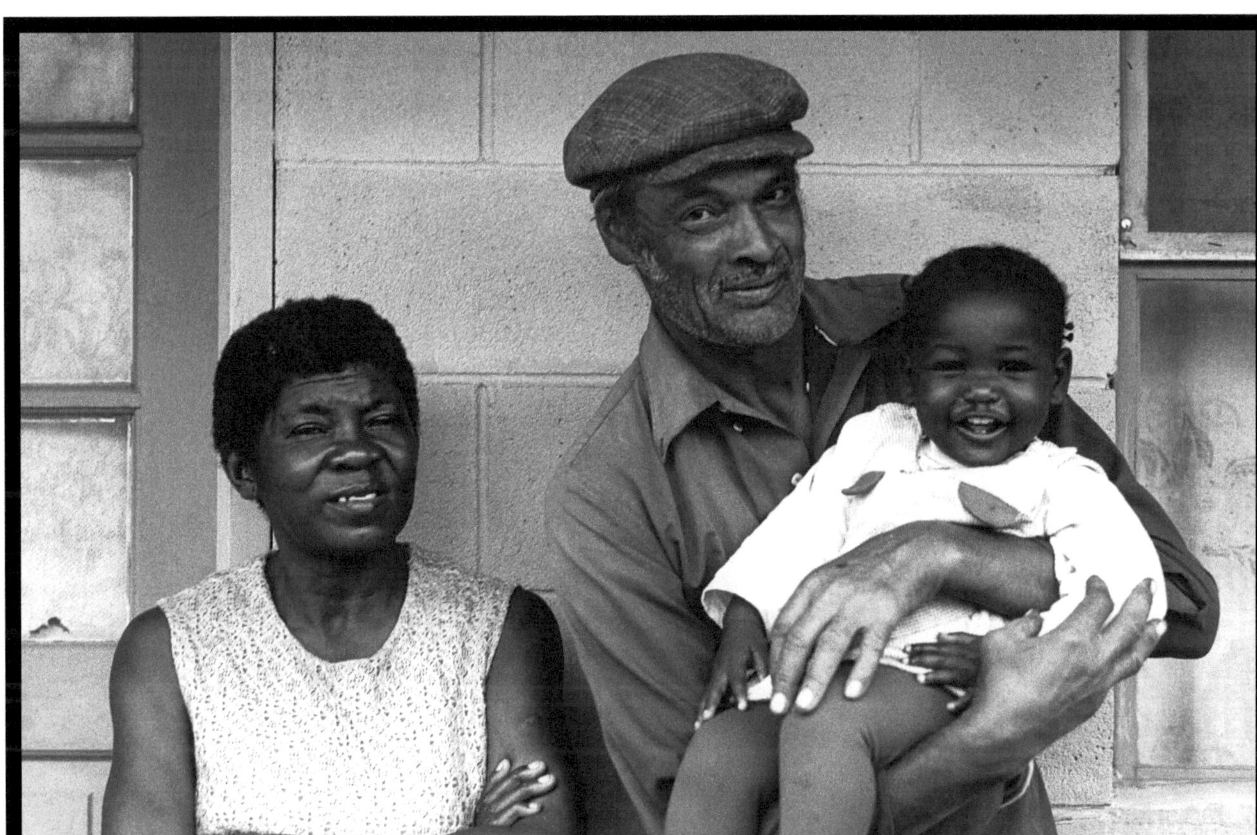

Tom Policano

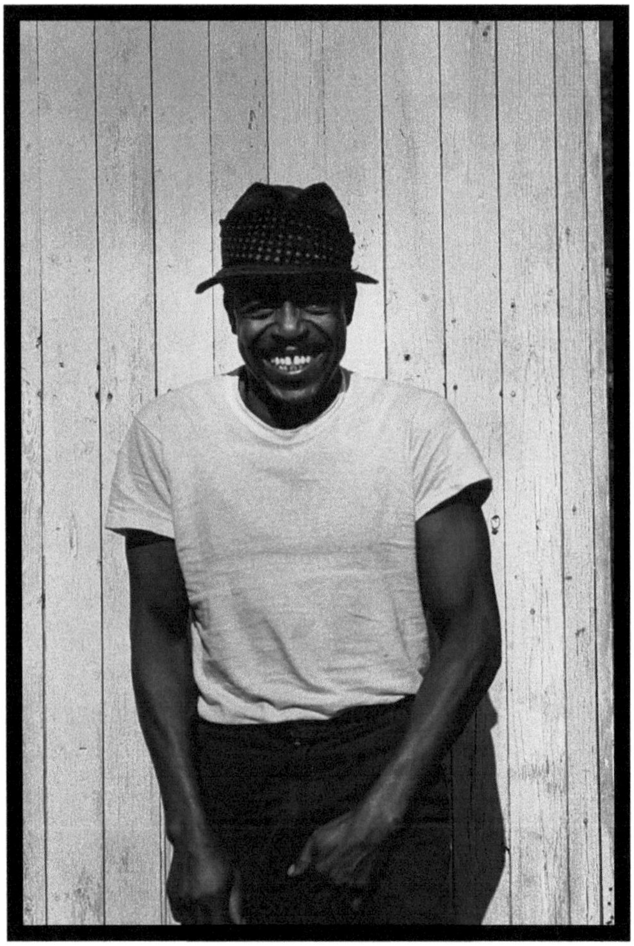

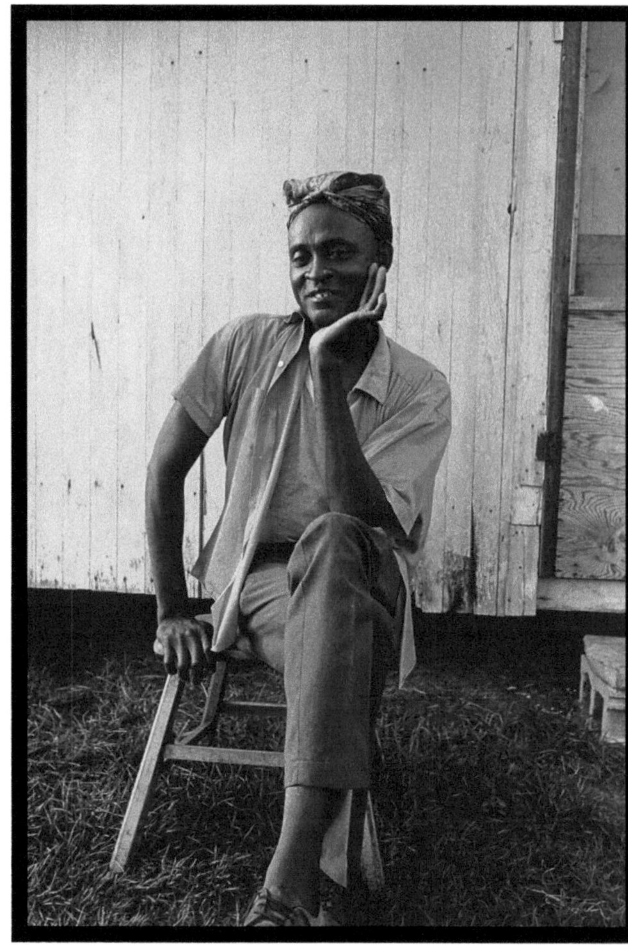

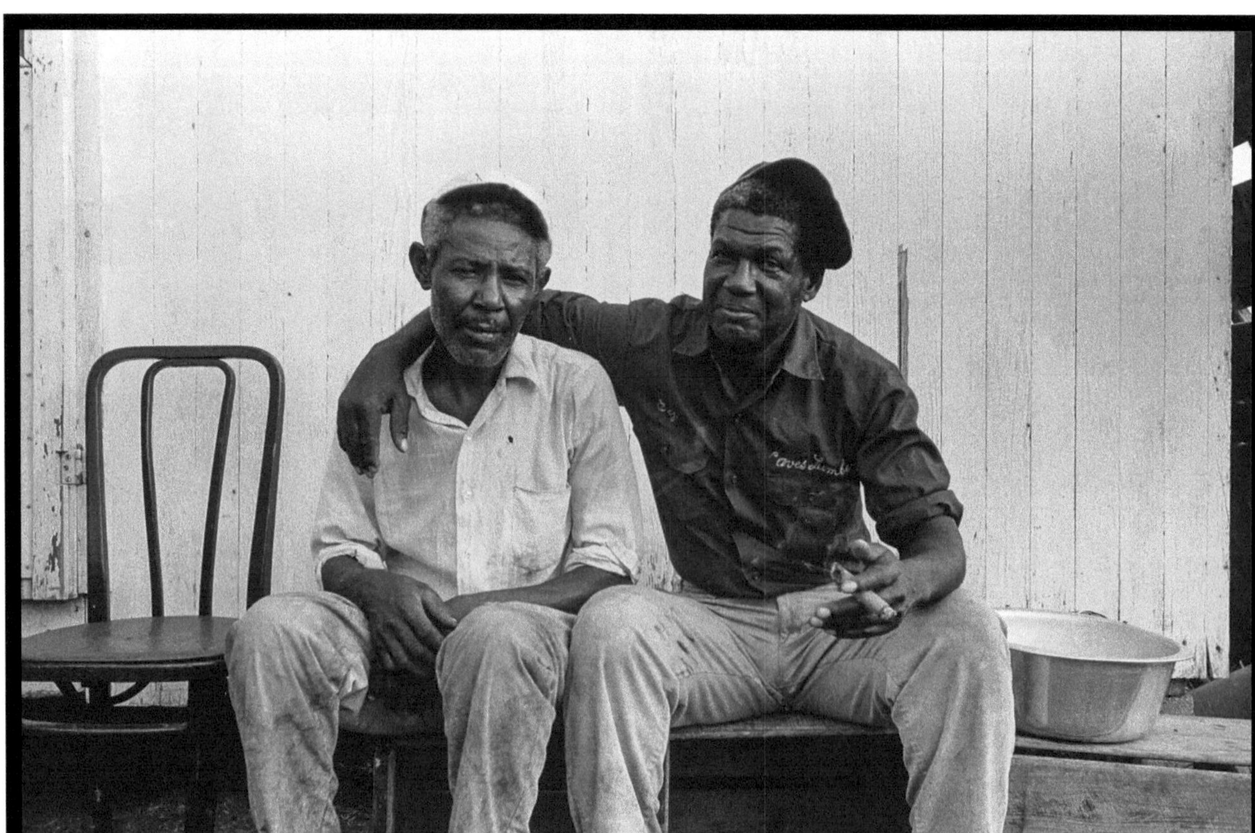

Slim Pickins

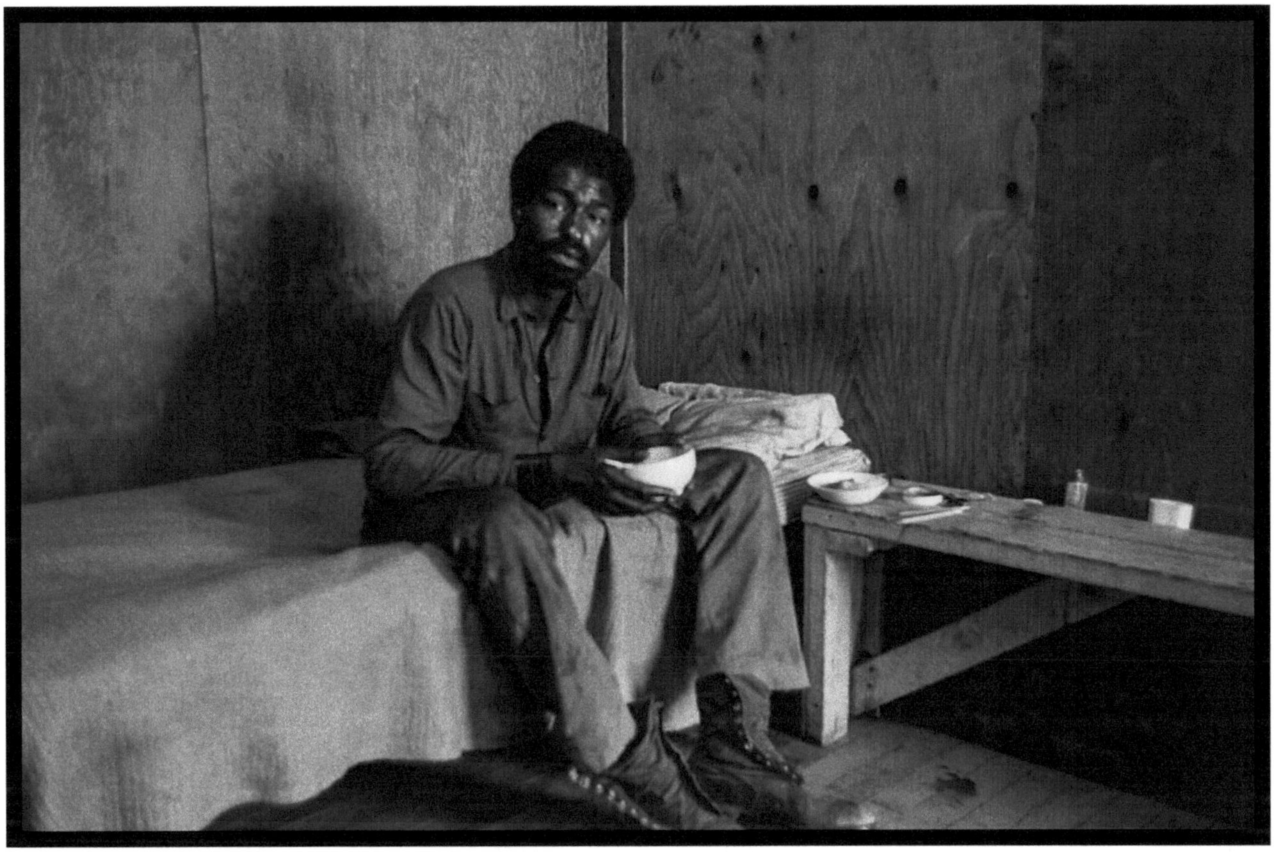

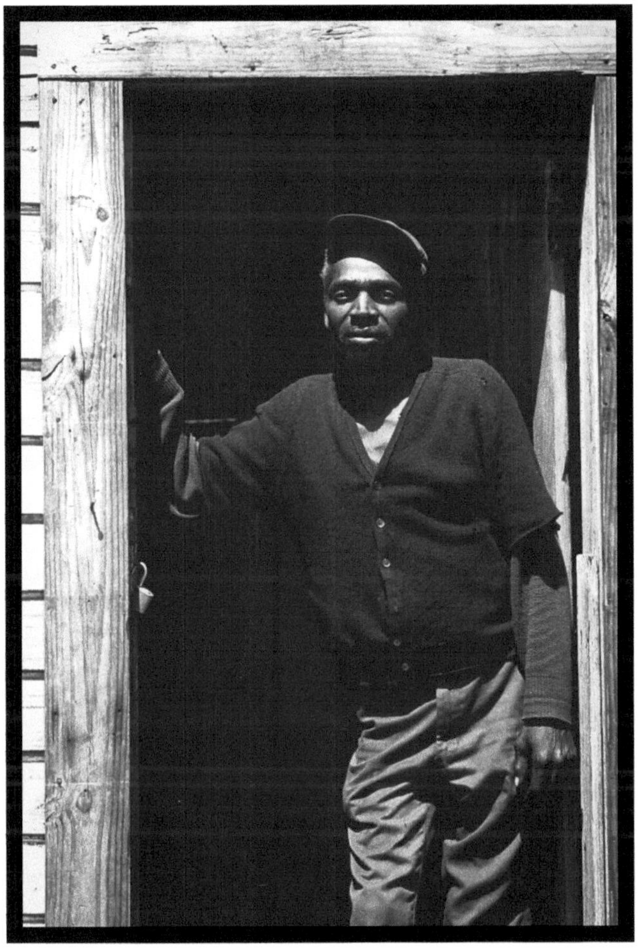

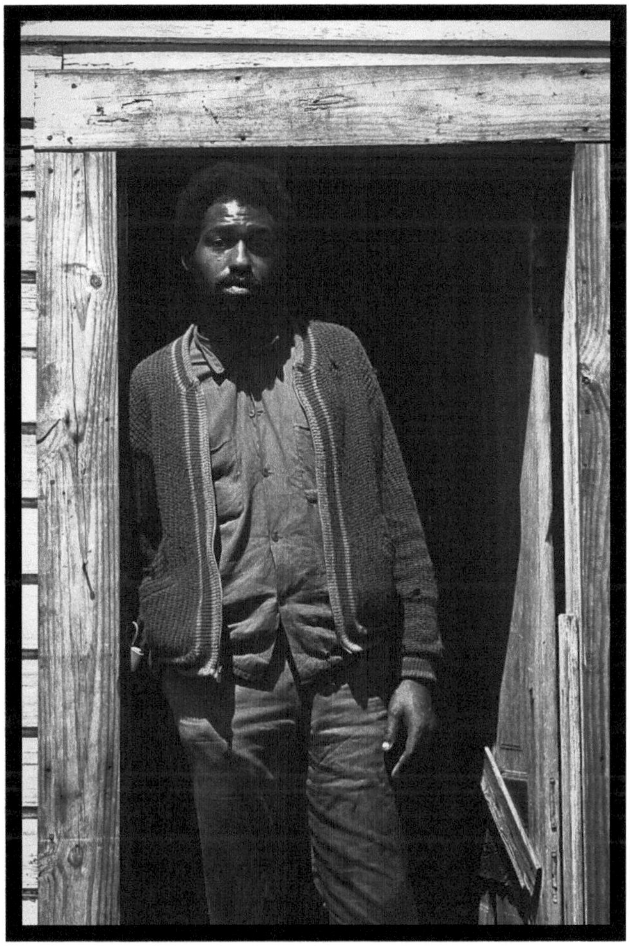

Tom Policano

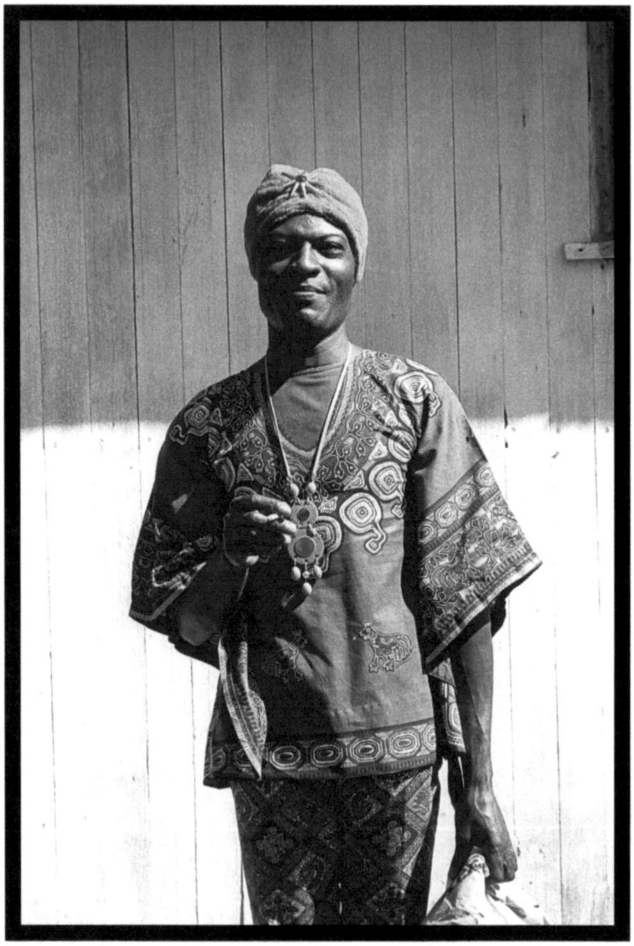

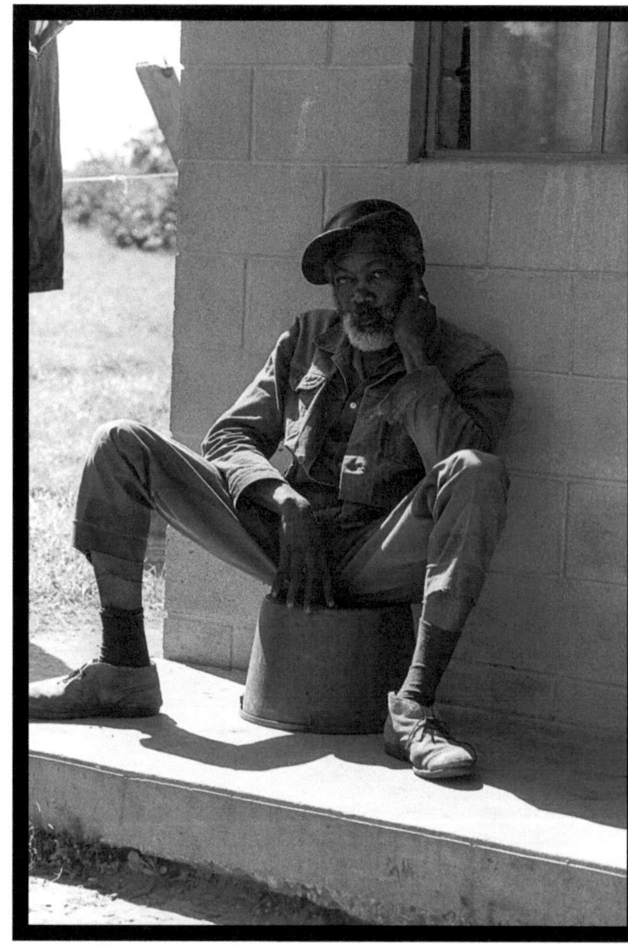

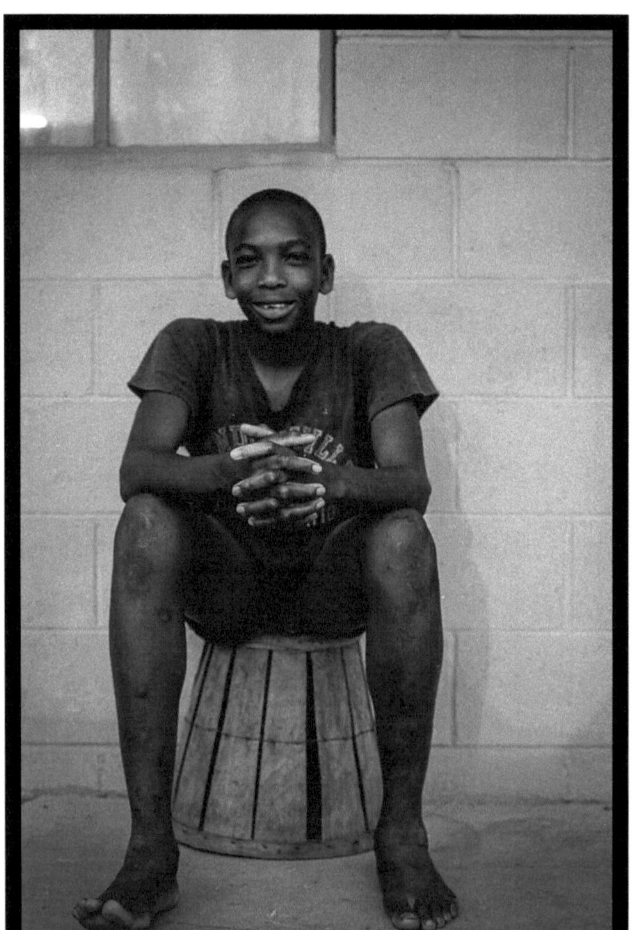

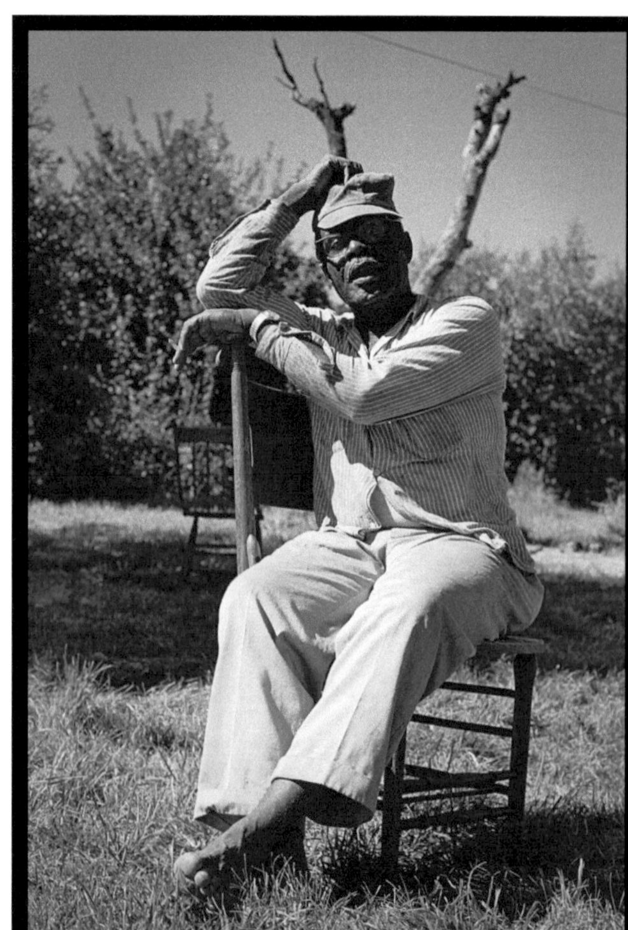

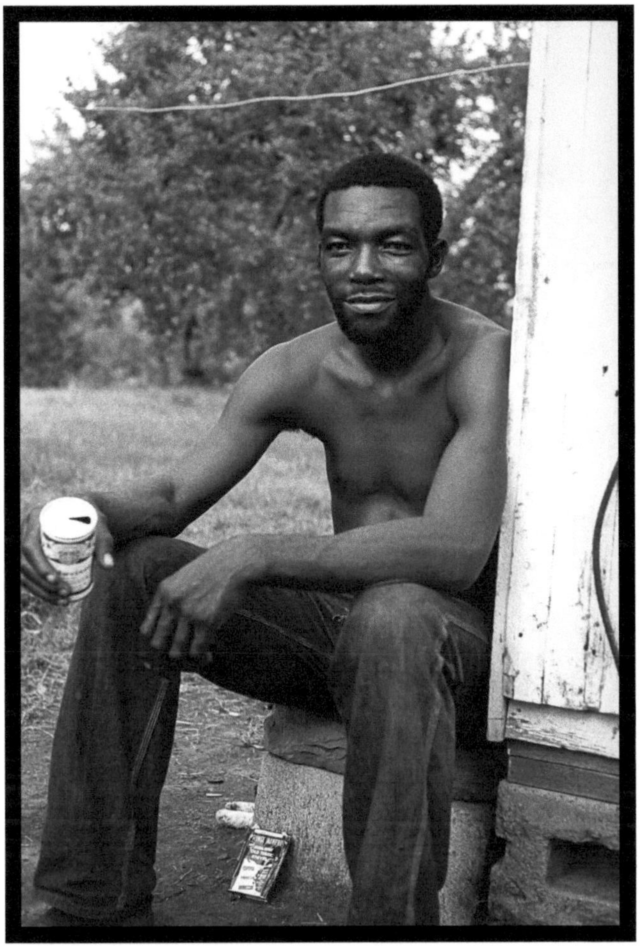

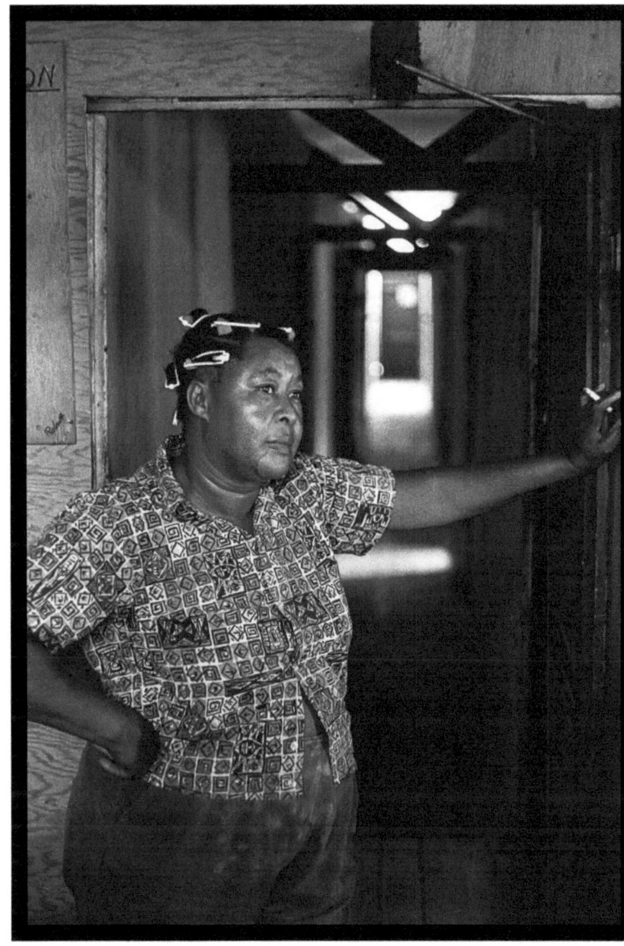

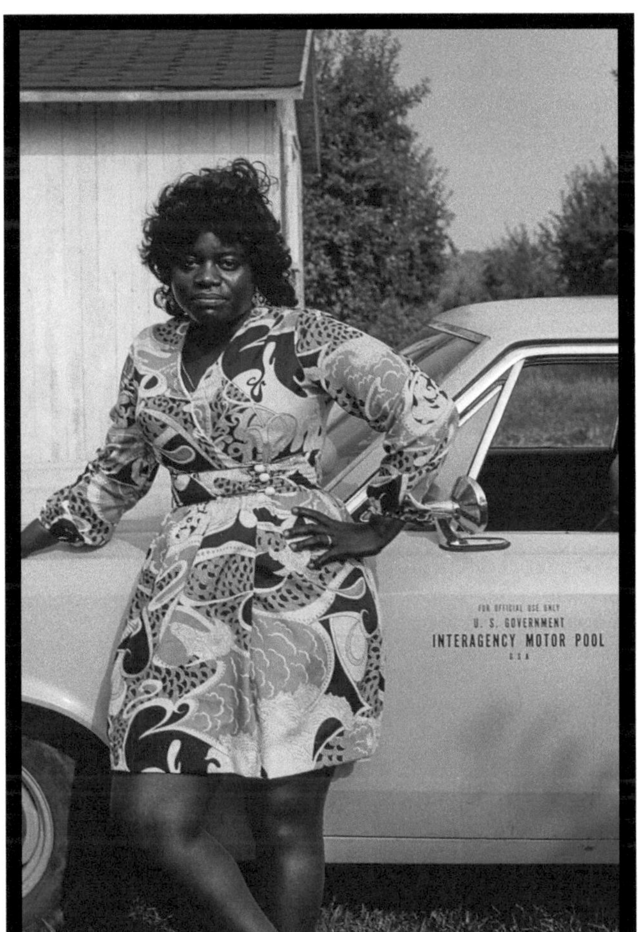

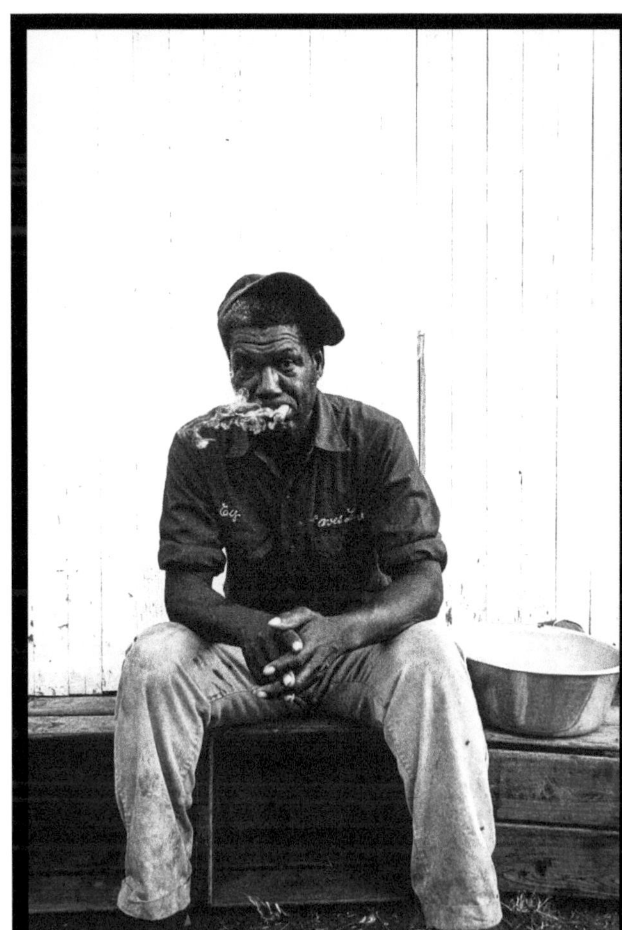

Tom Policano

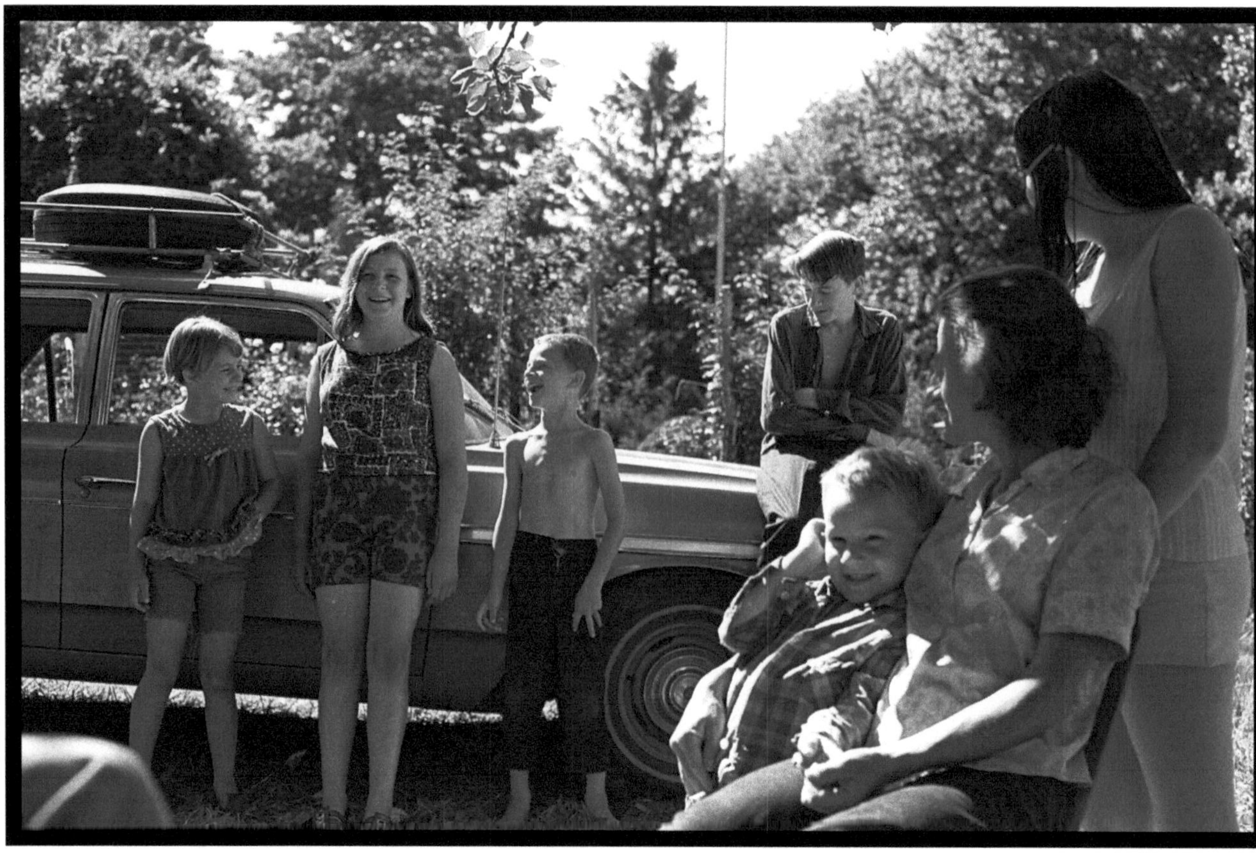

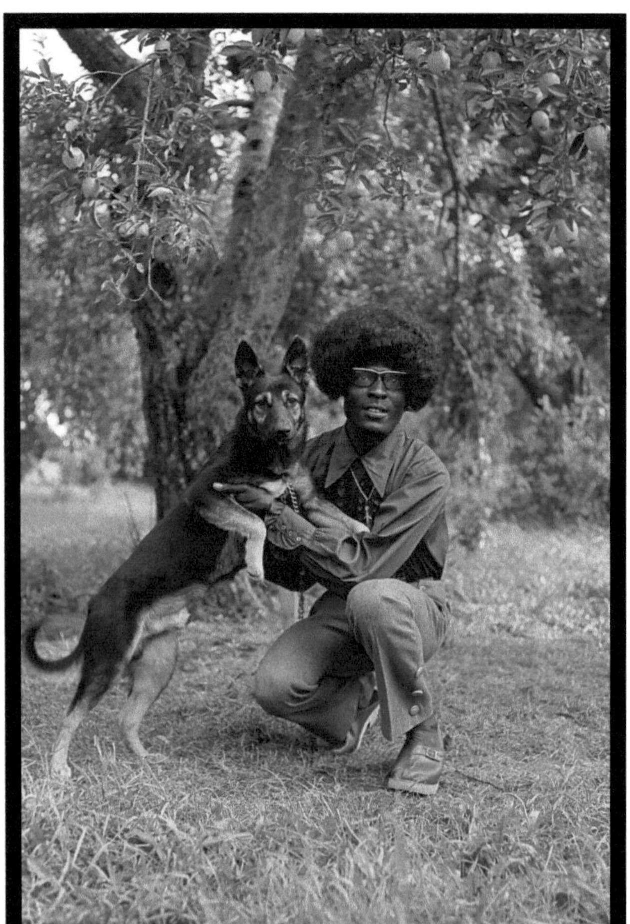

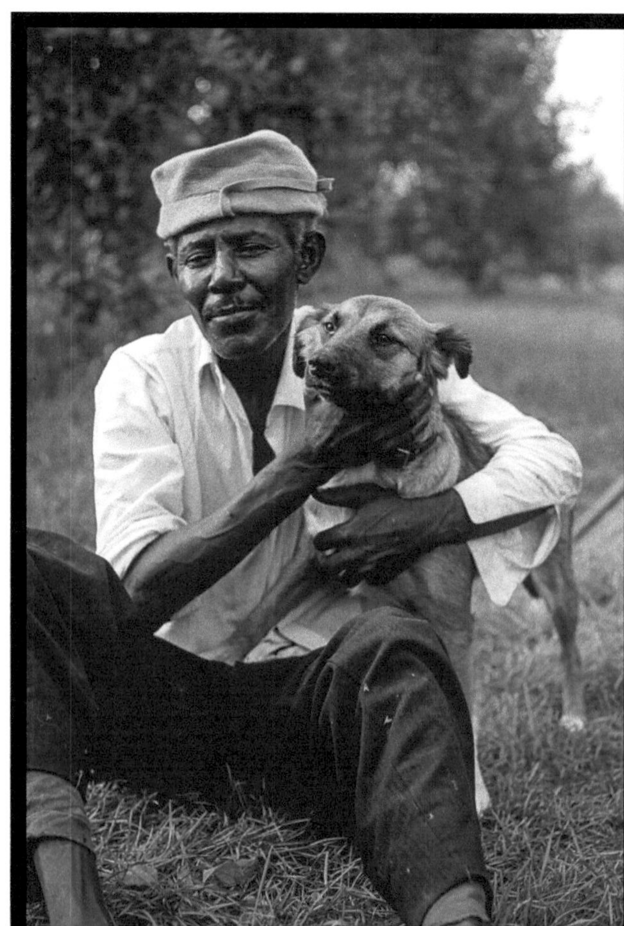

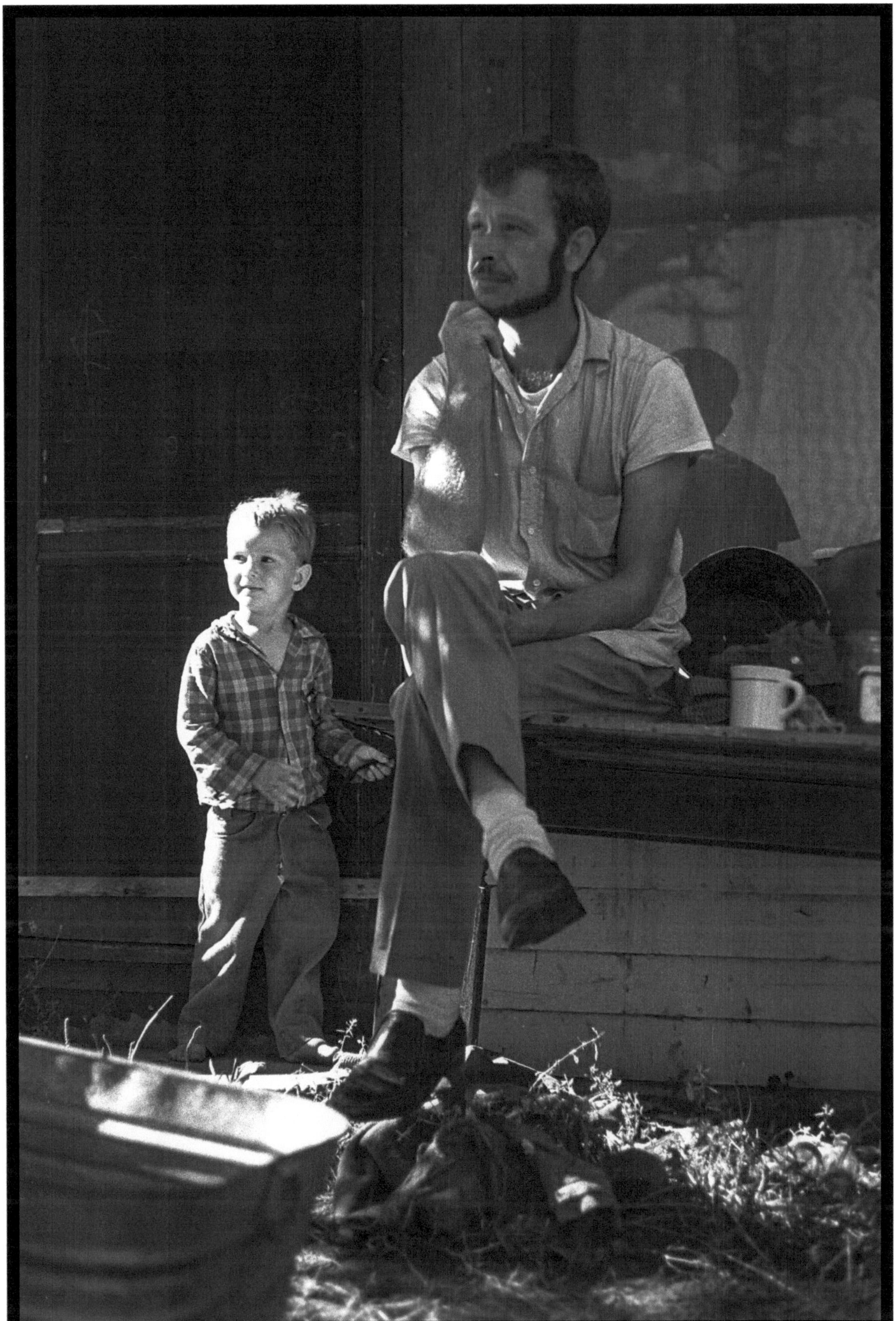

Tom Policano

South — At Work

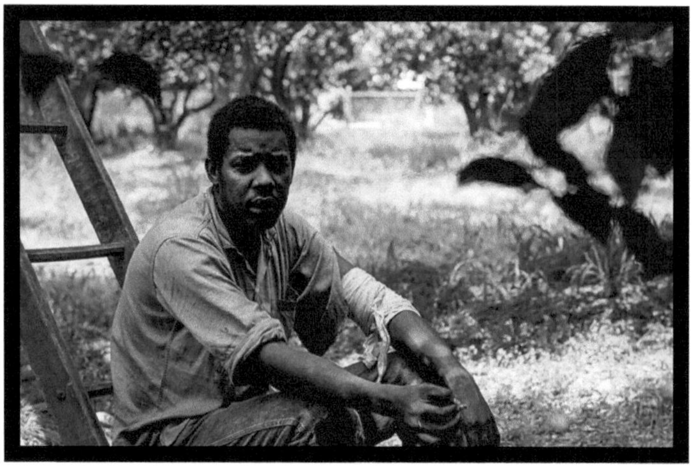

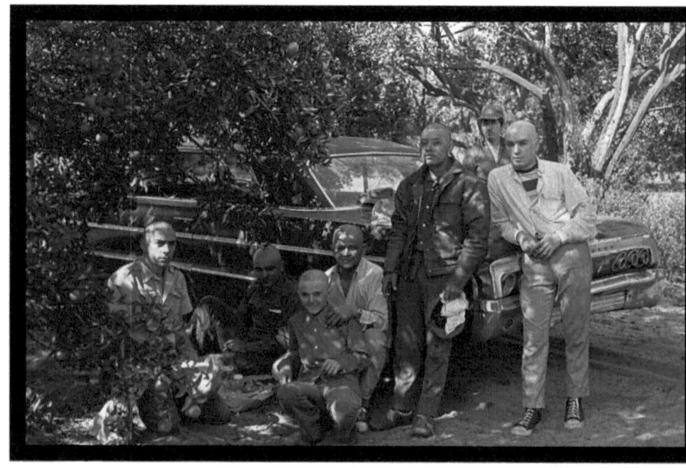

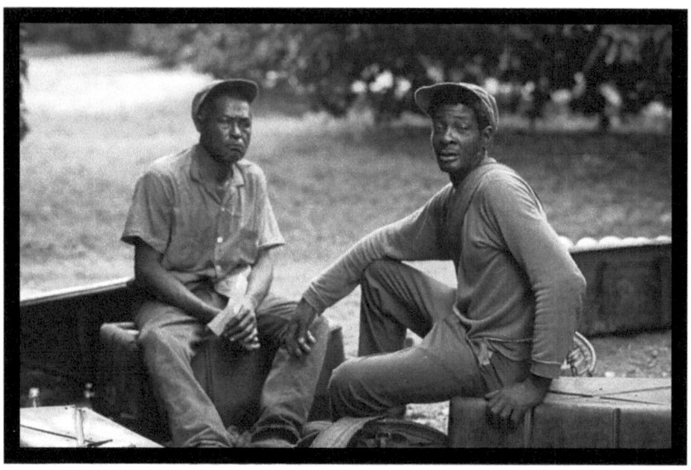

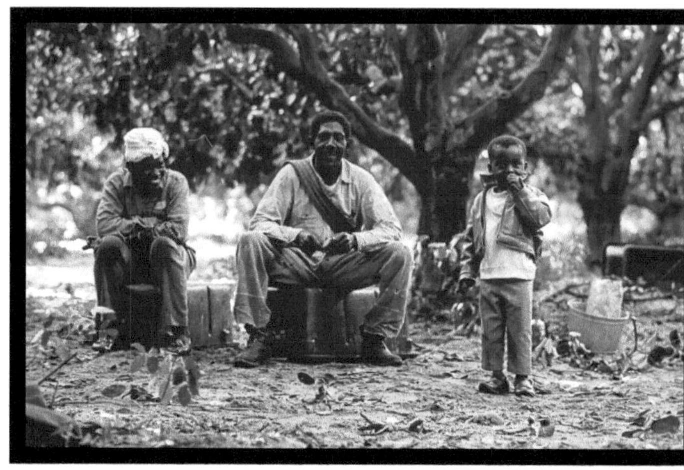

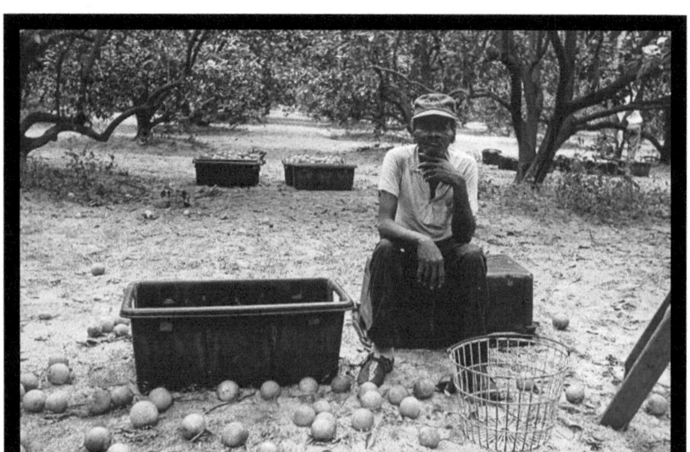

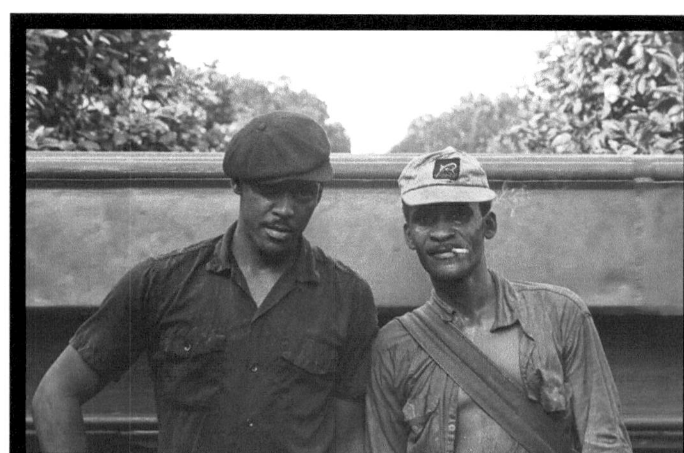

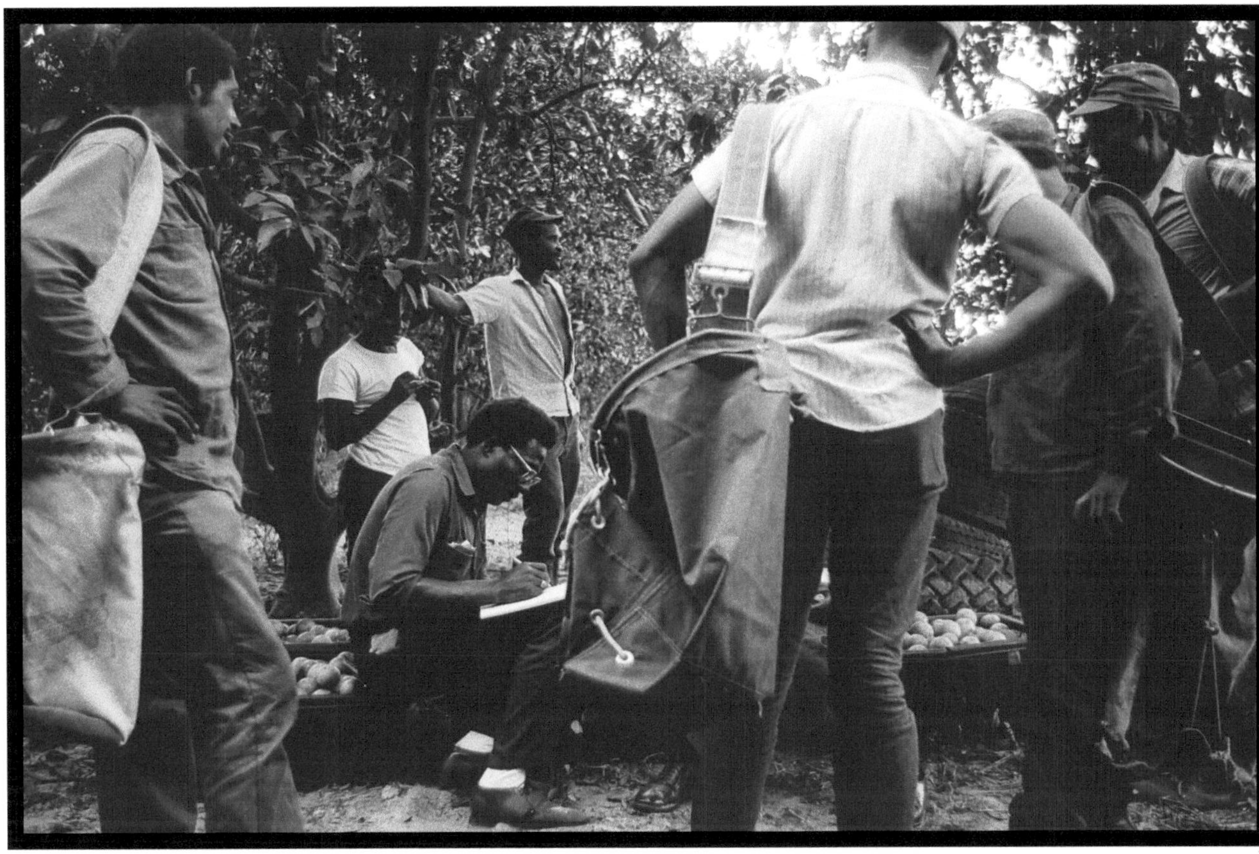

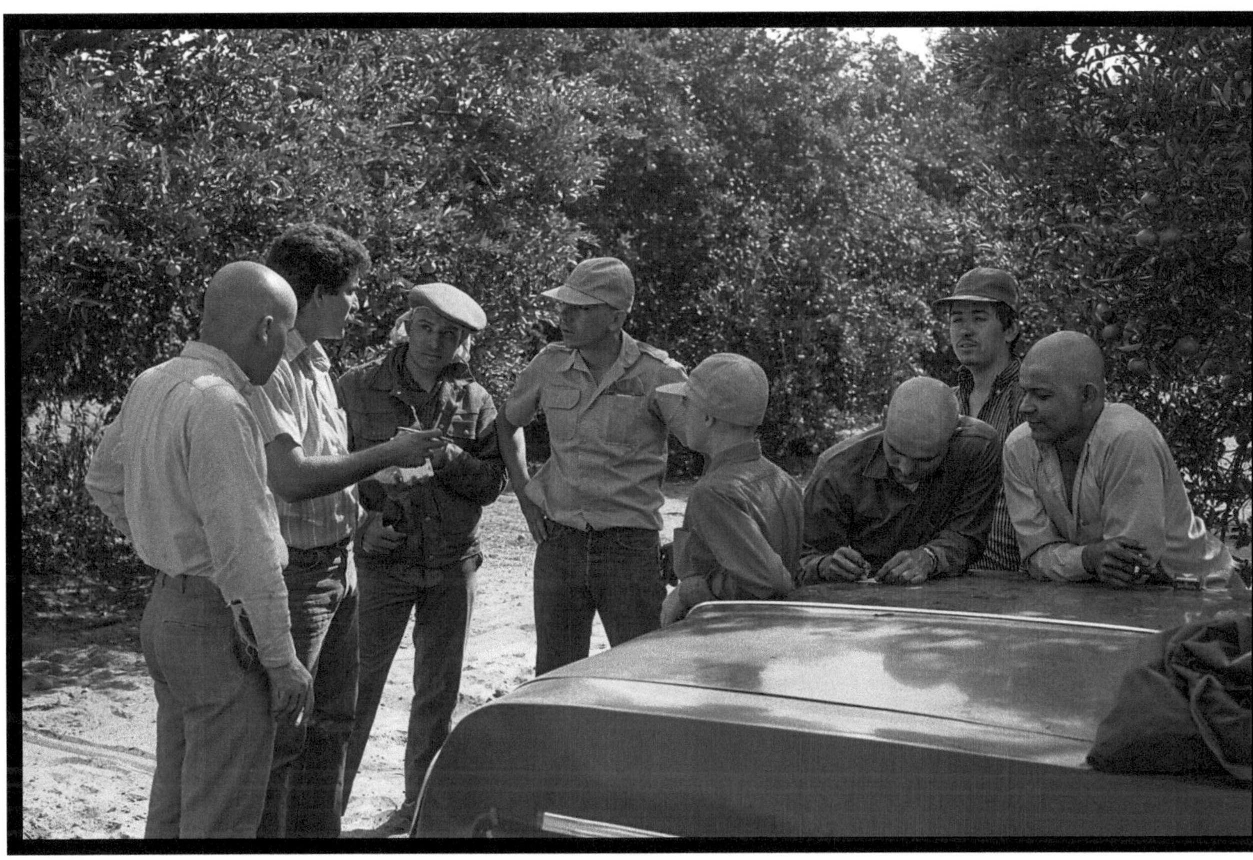

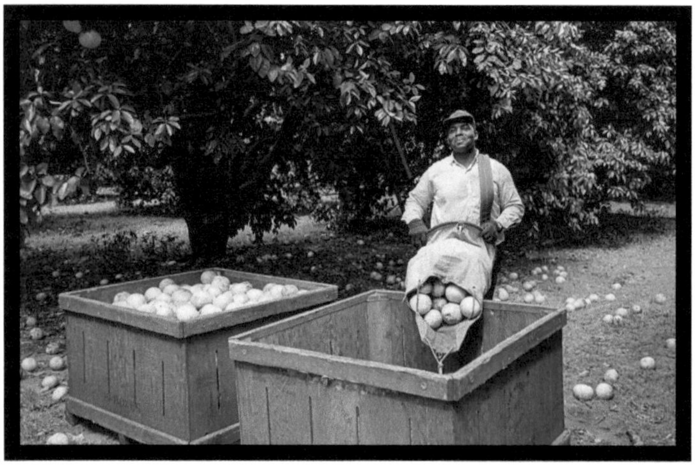
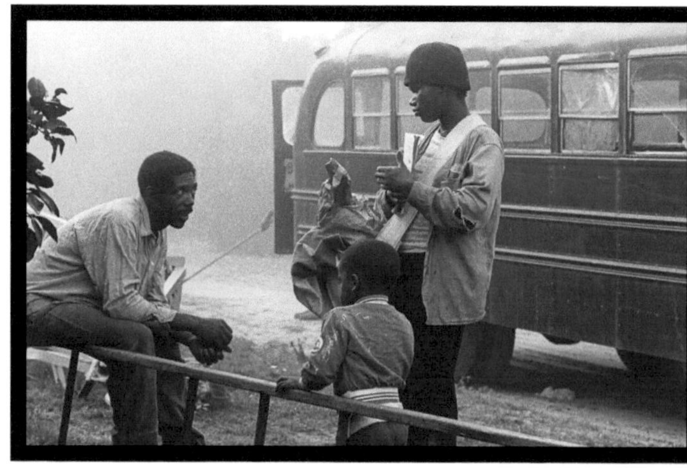
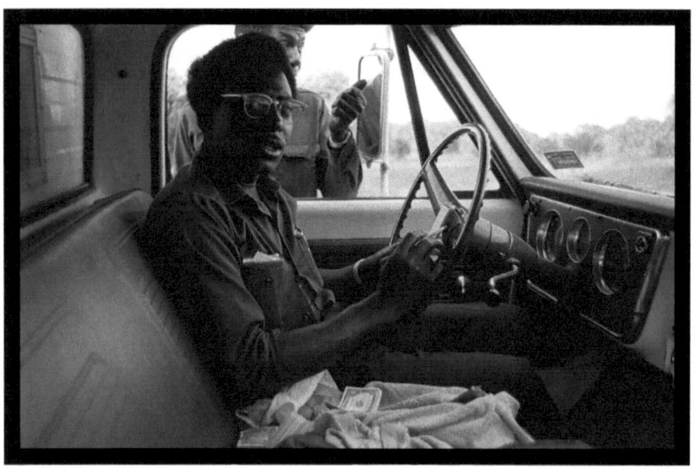
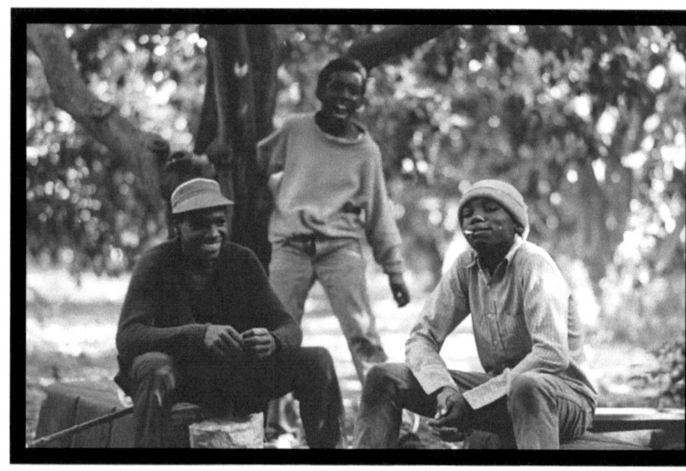
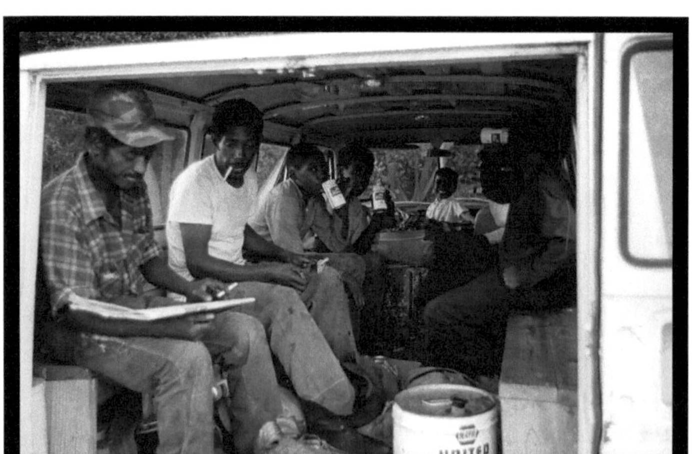
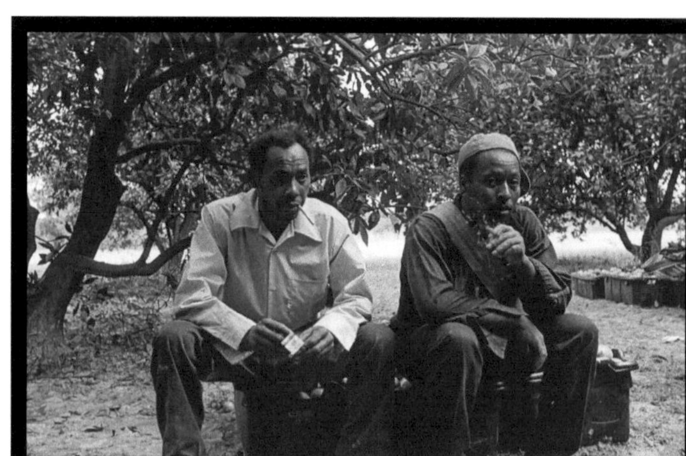

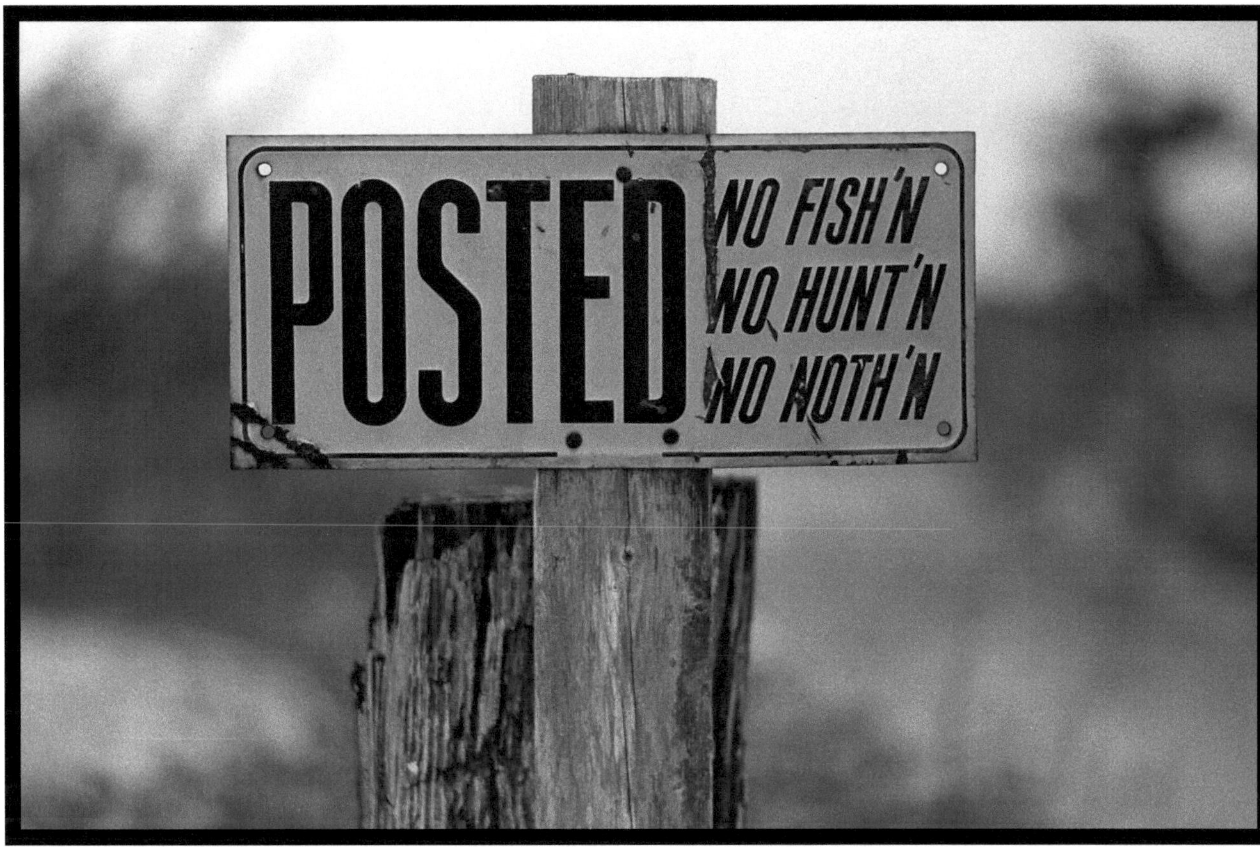

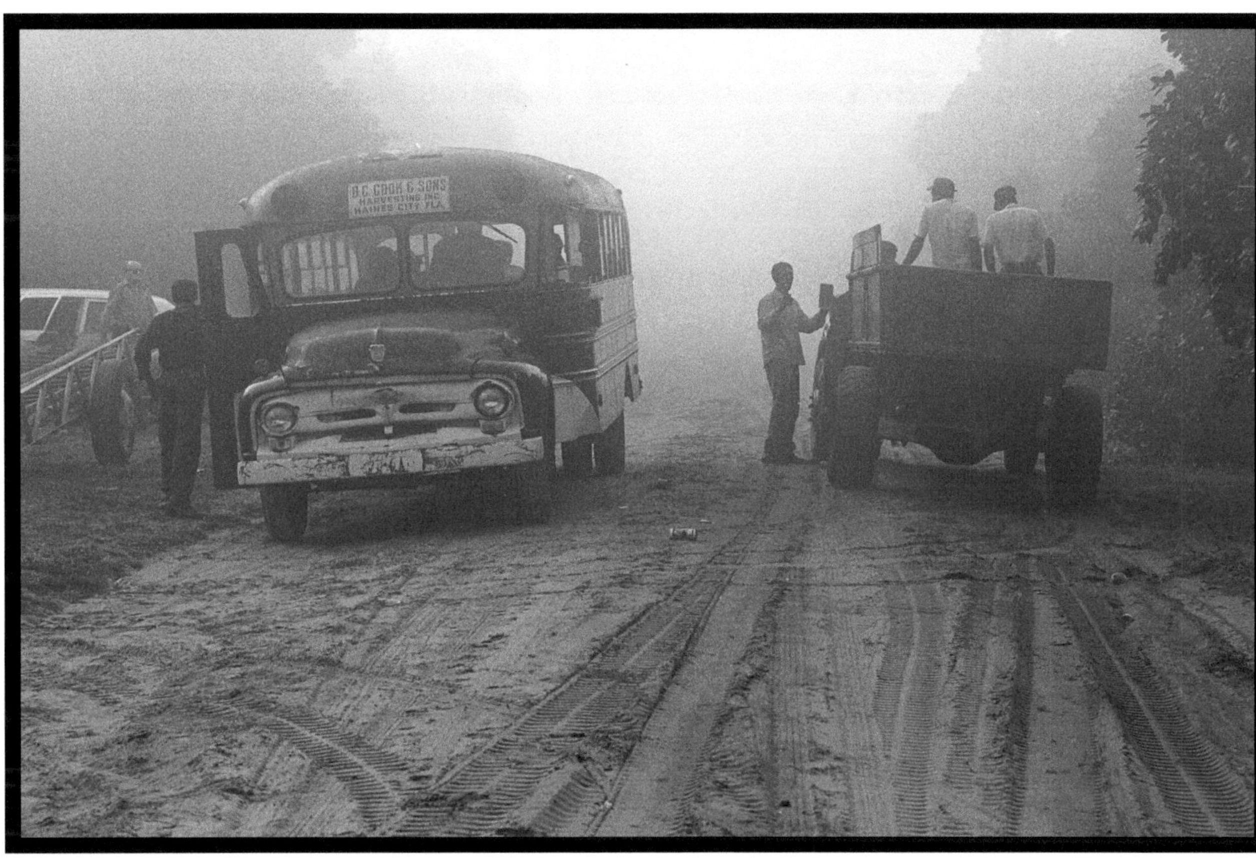

Tom Policano

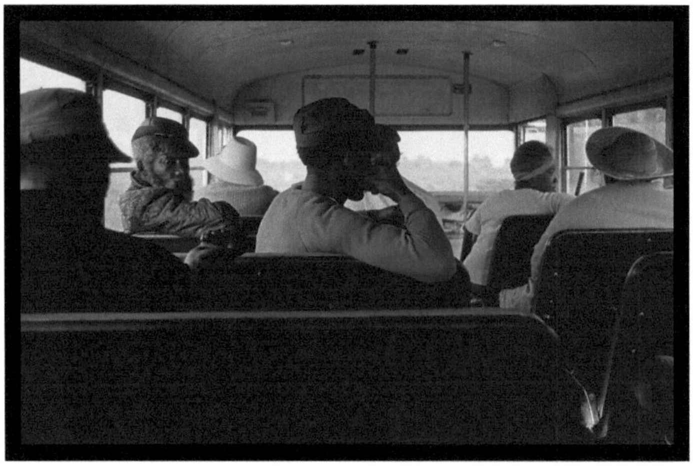
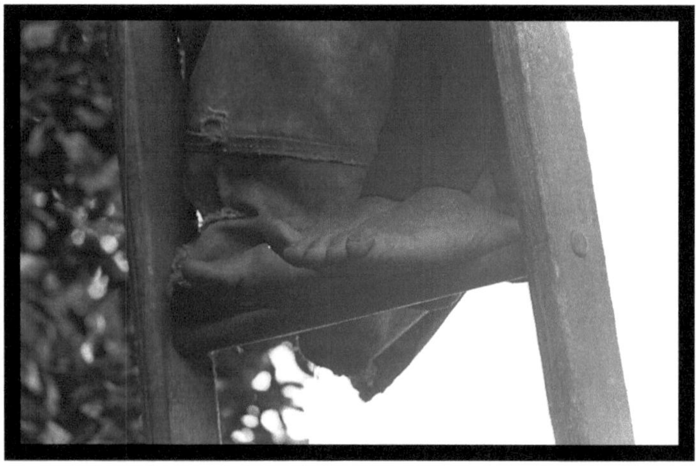
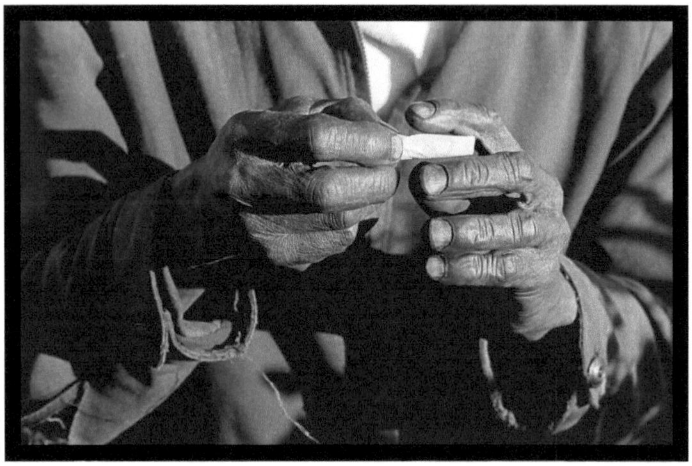
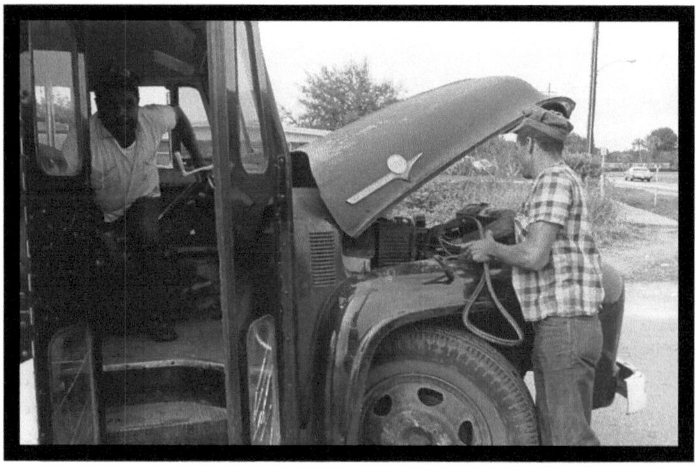
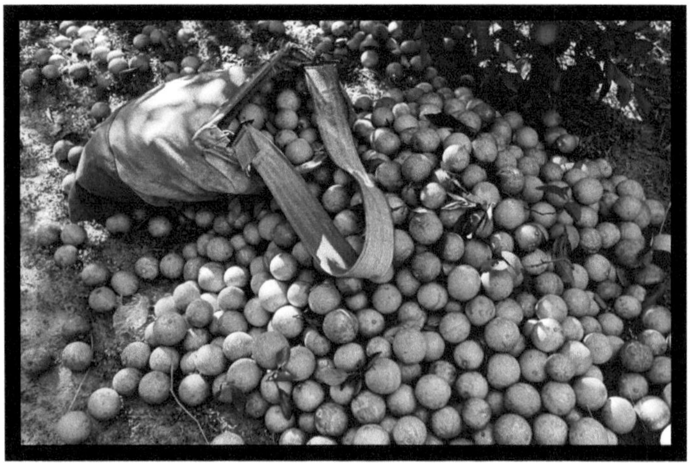
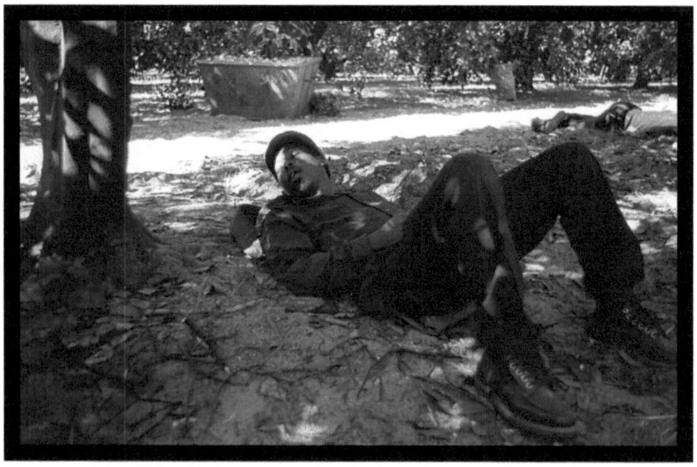

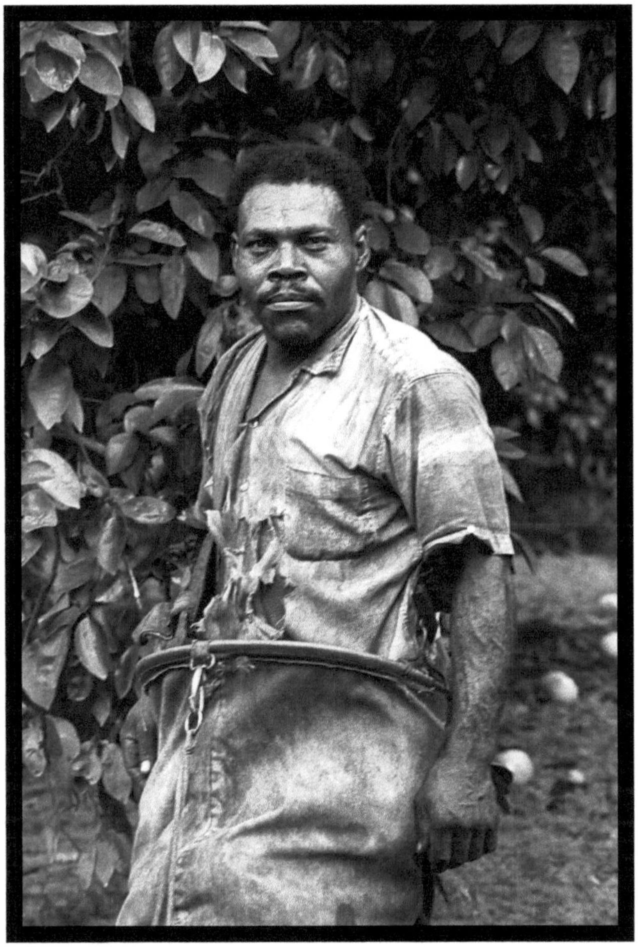

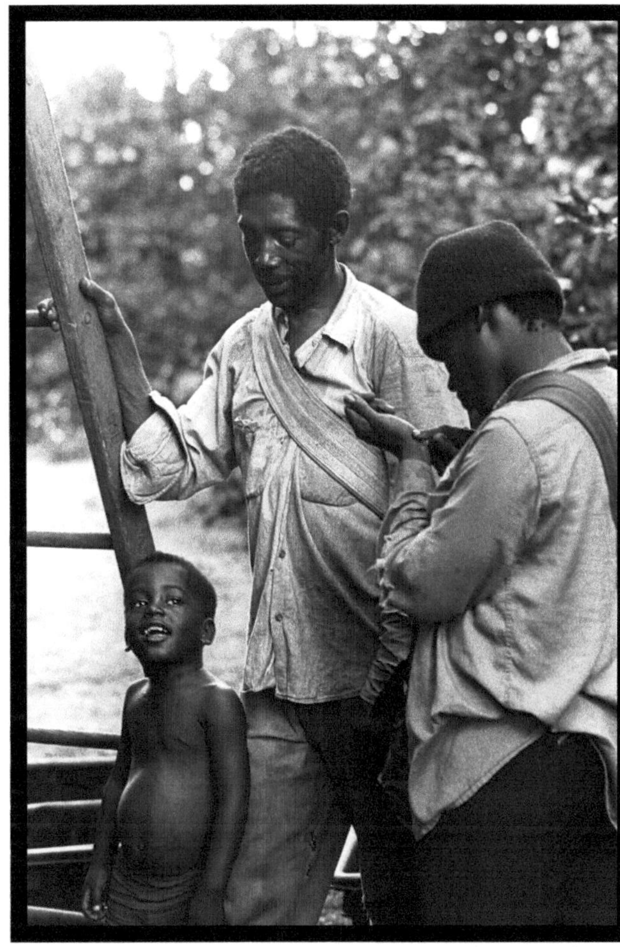

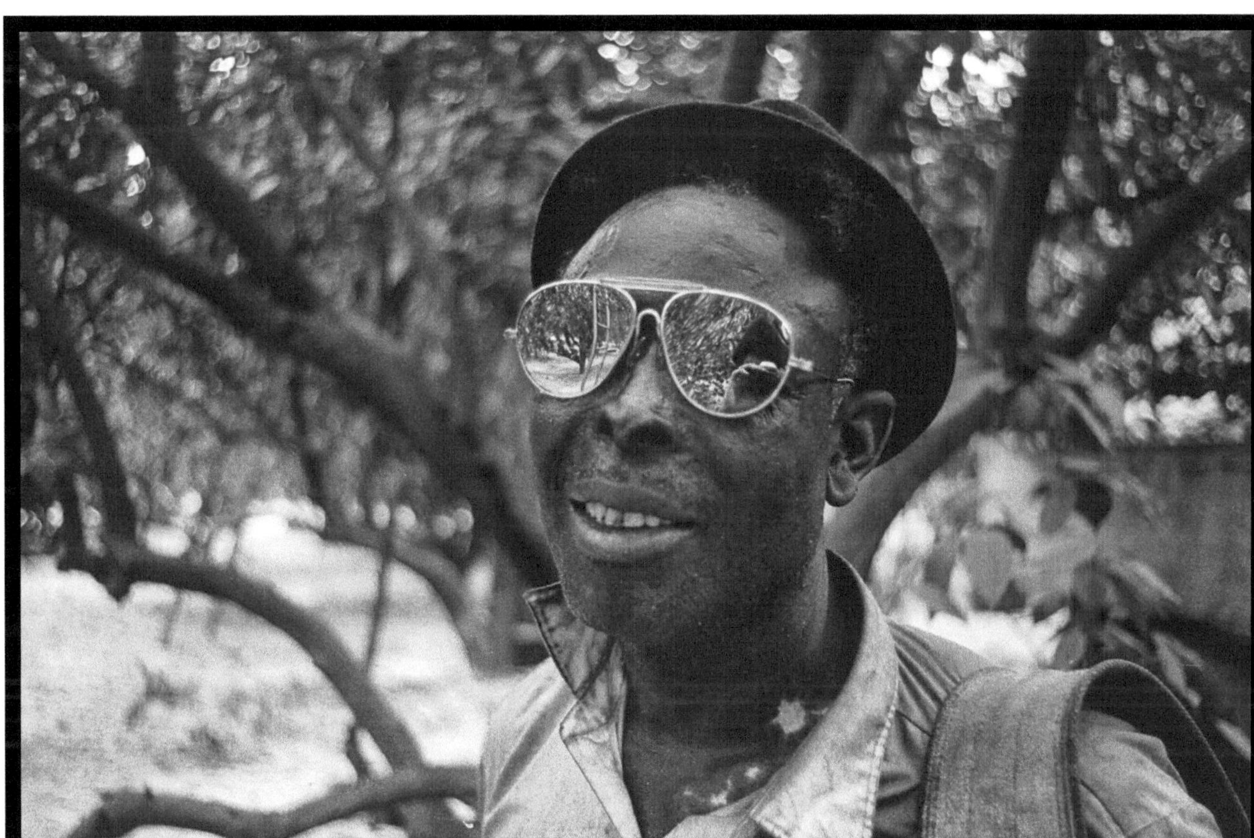

Tom Policano

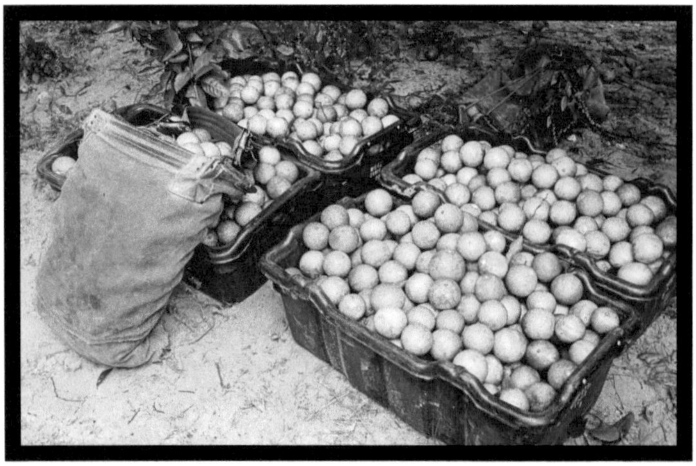
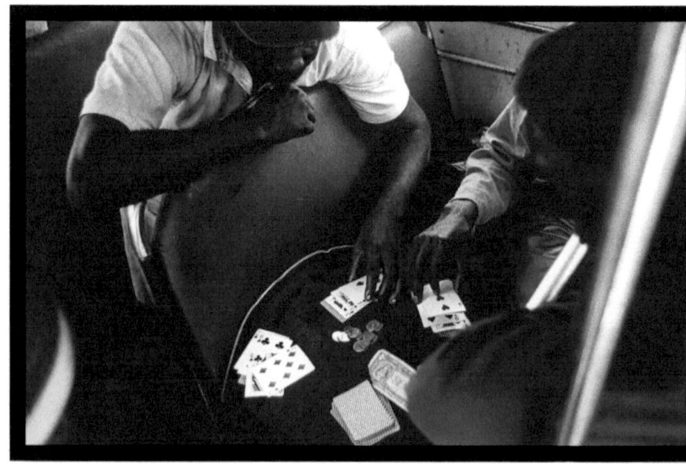
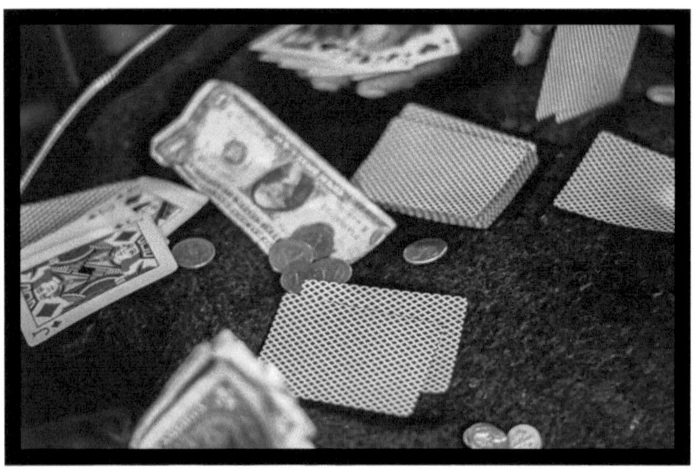
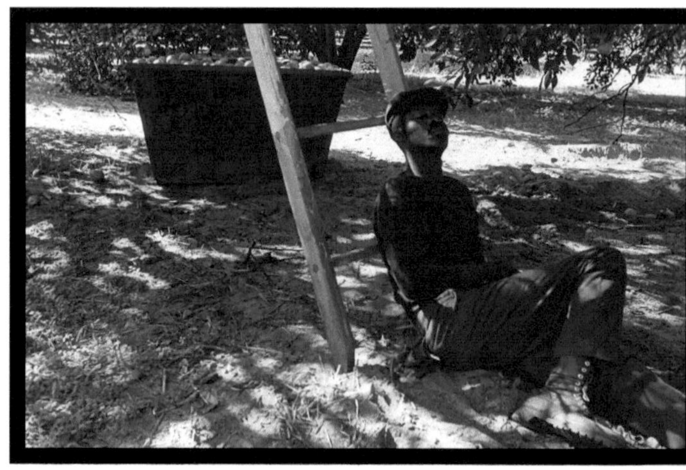
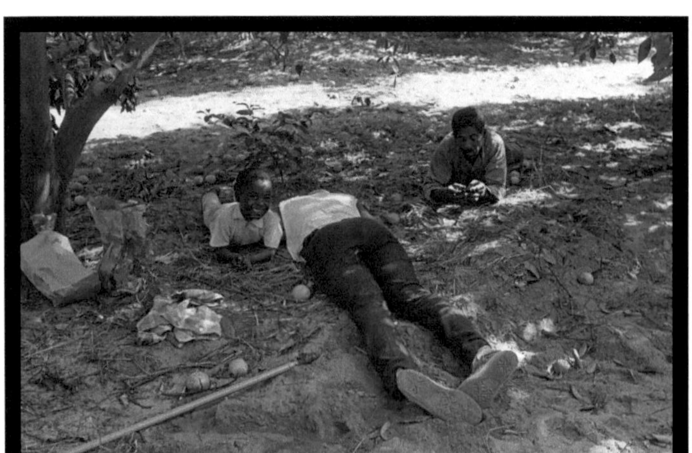
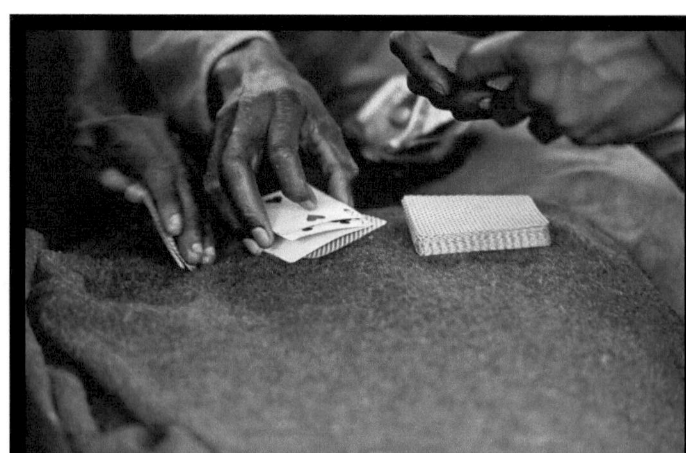

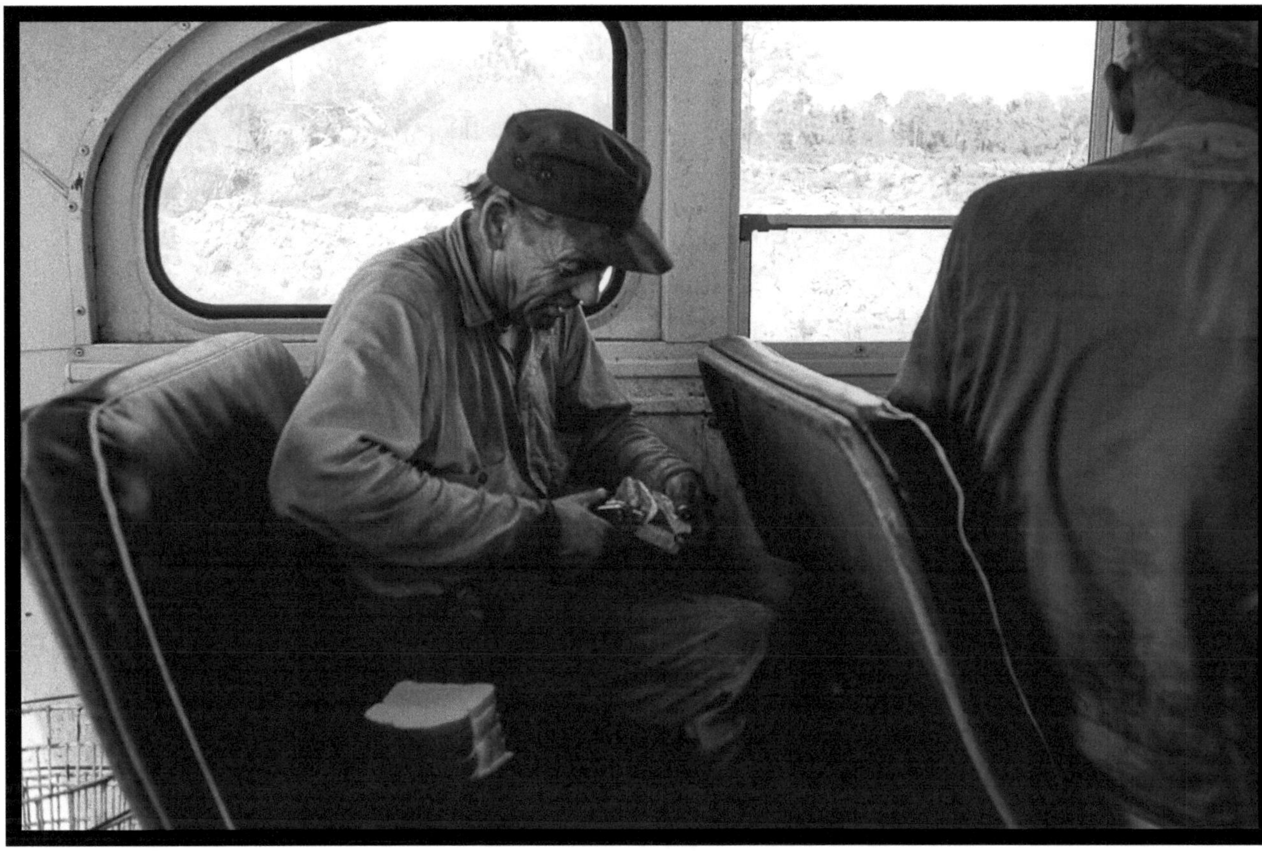

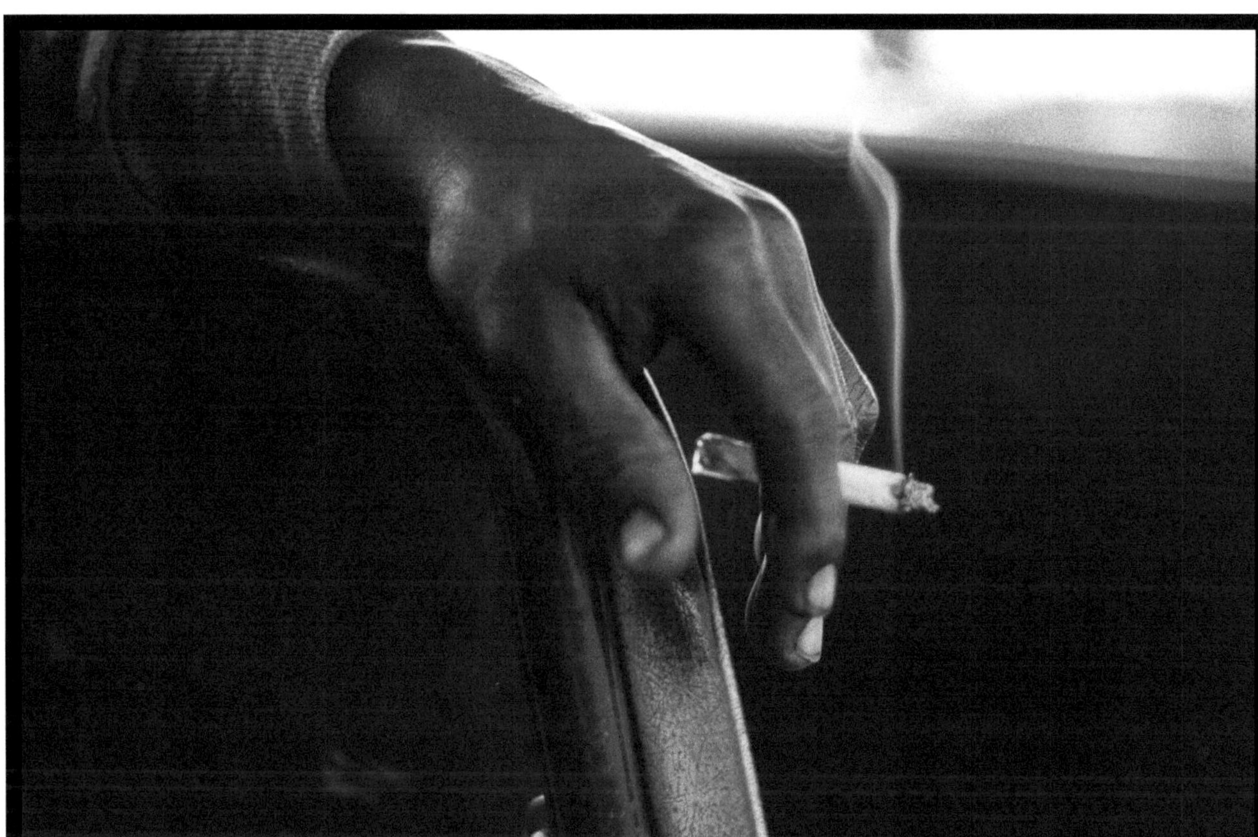

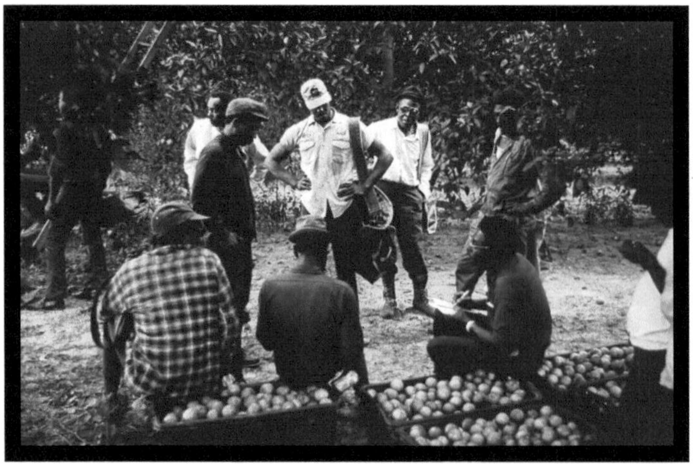

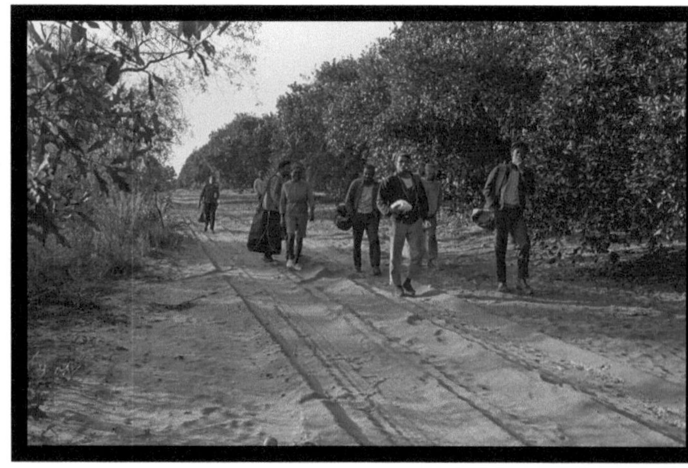

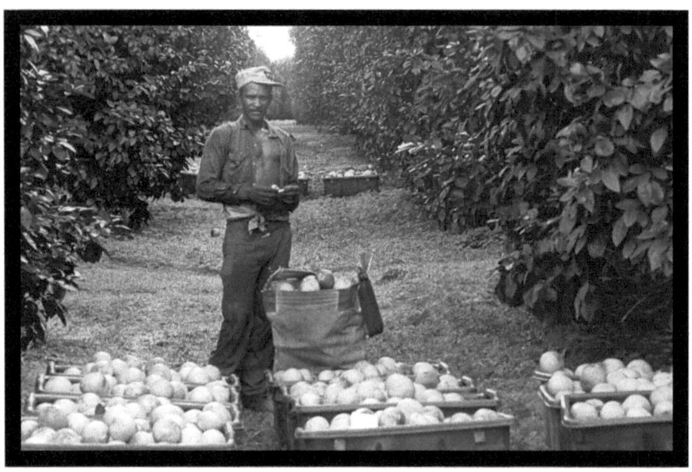

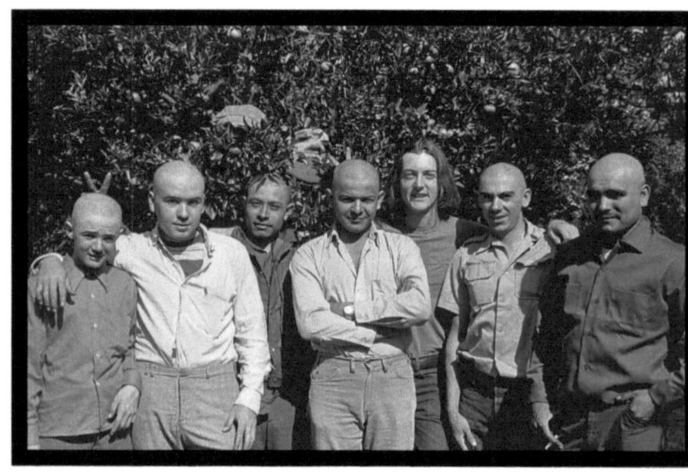

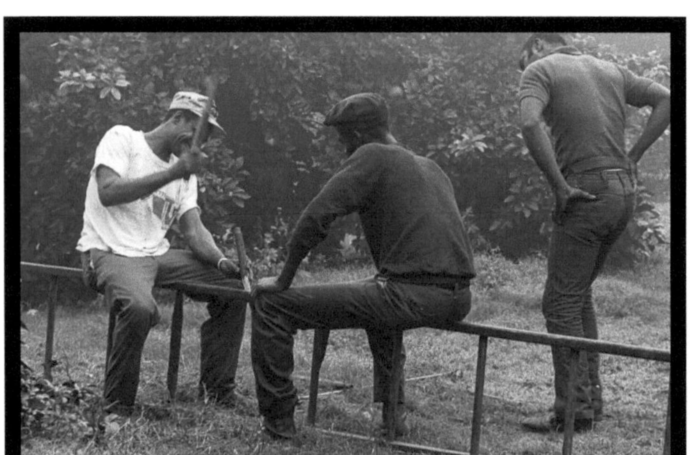

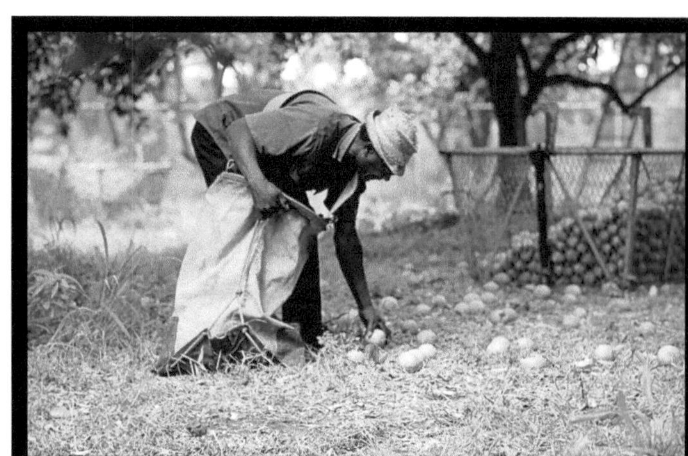

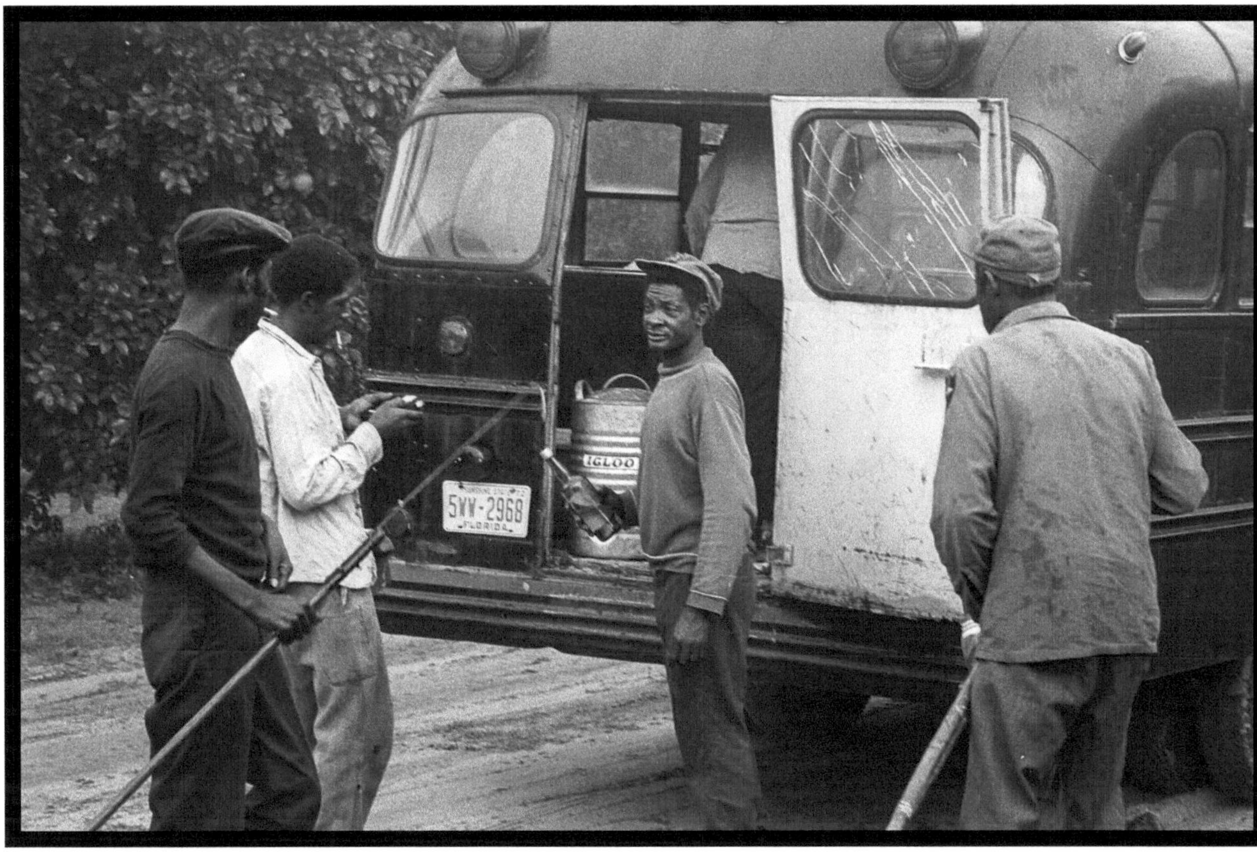
IGLOO
5WW-2968
FLORIDA

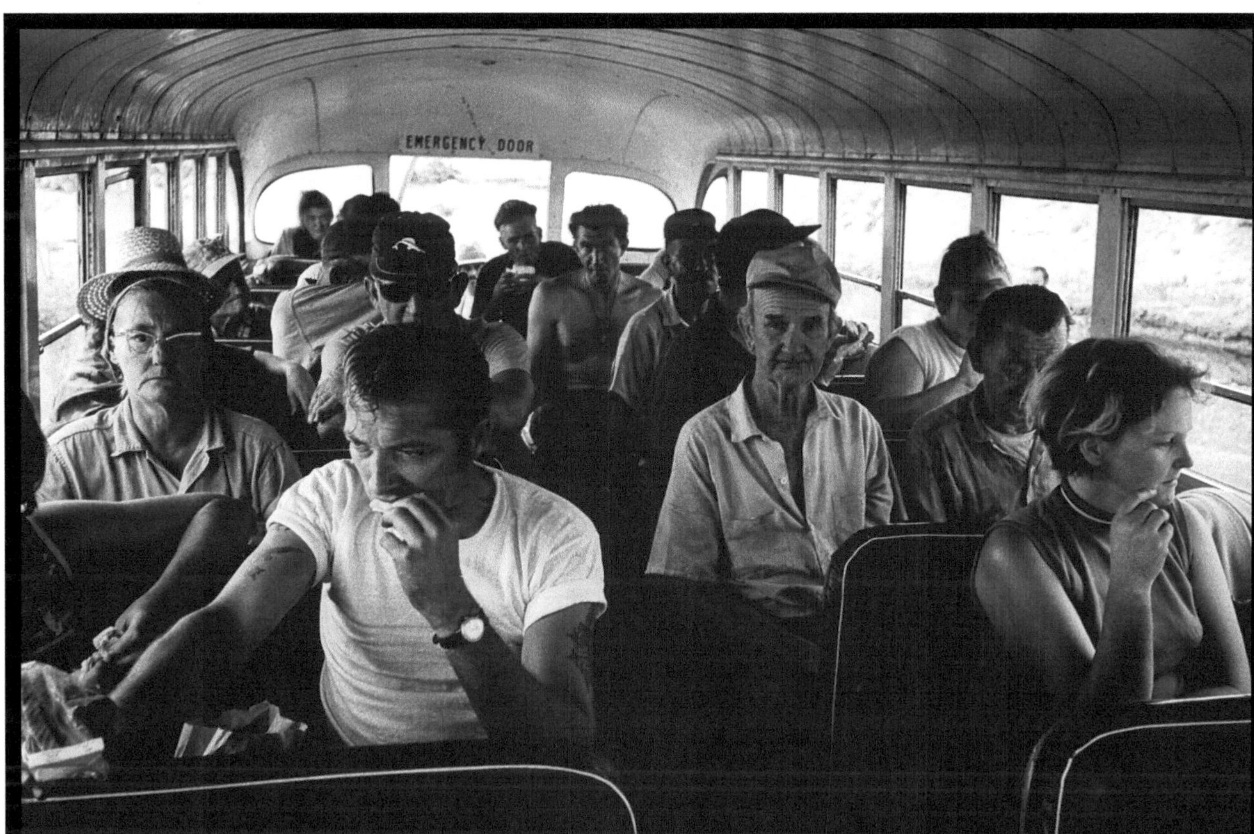
EMERGENCY DOOR

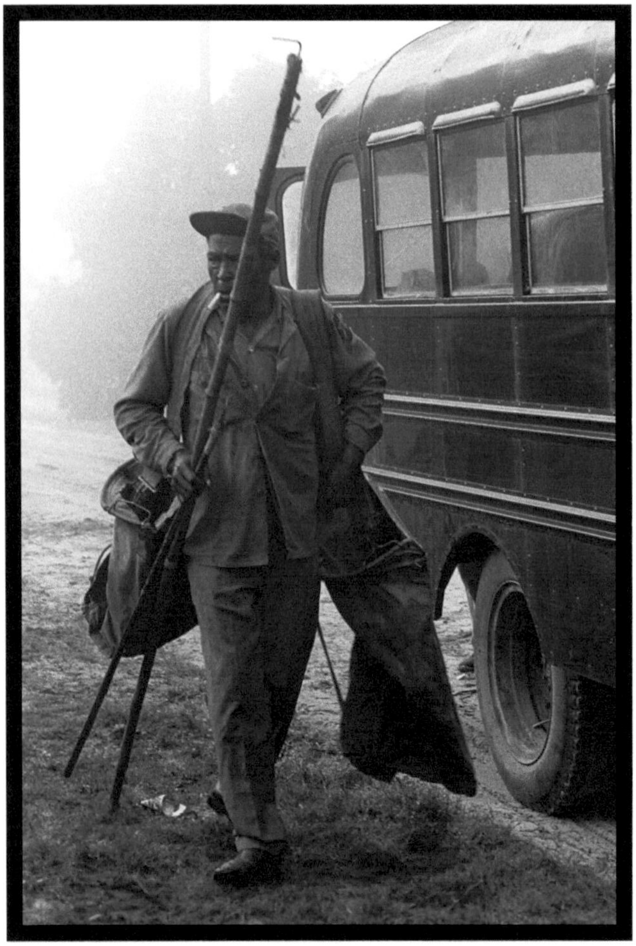

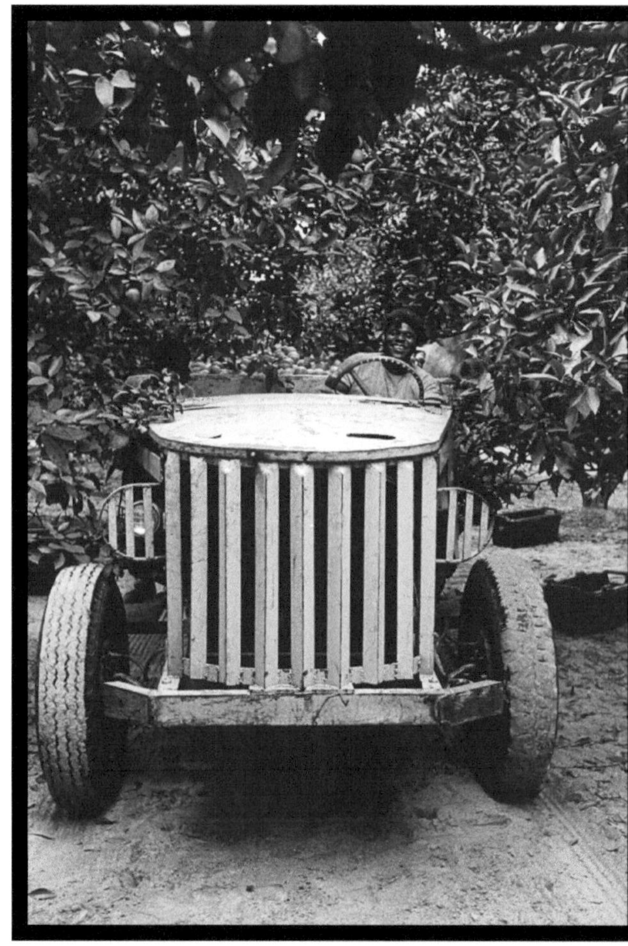

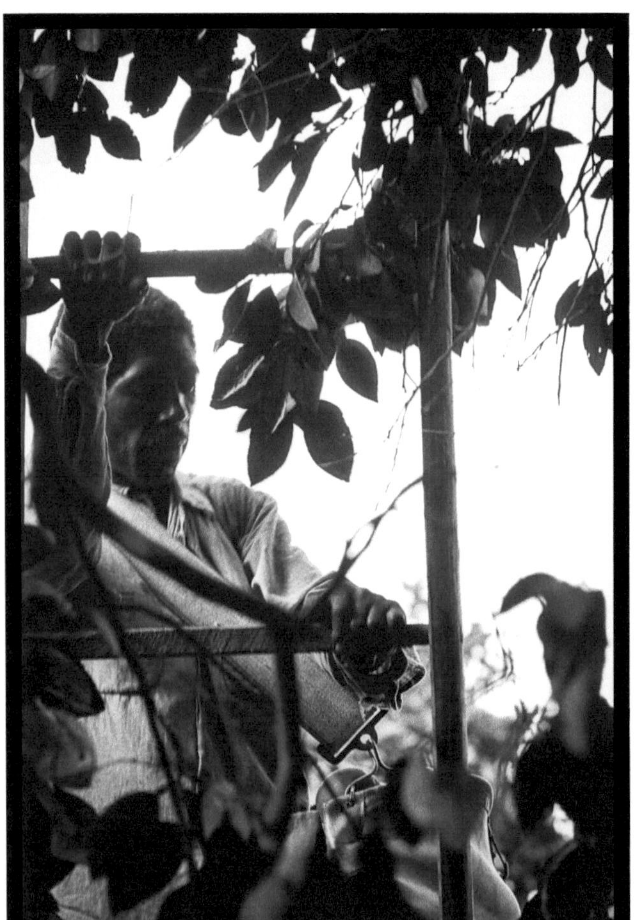

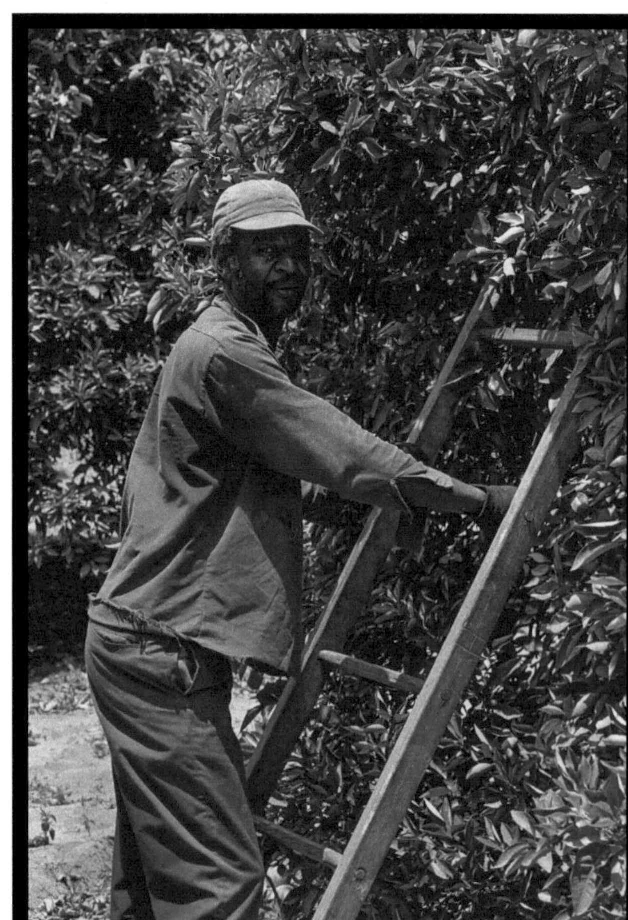

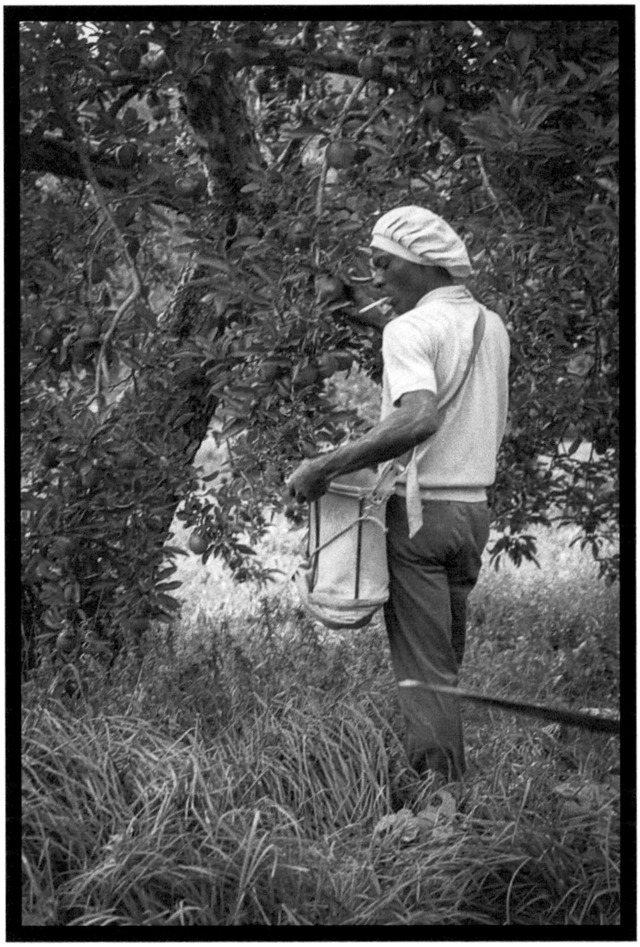

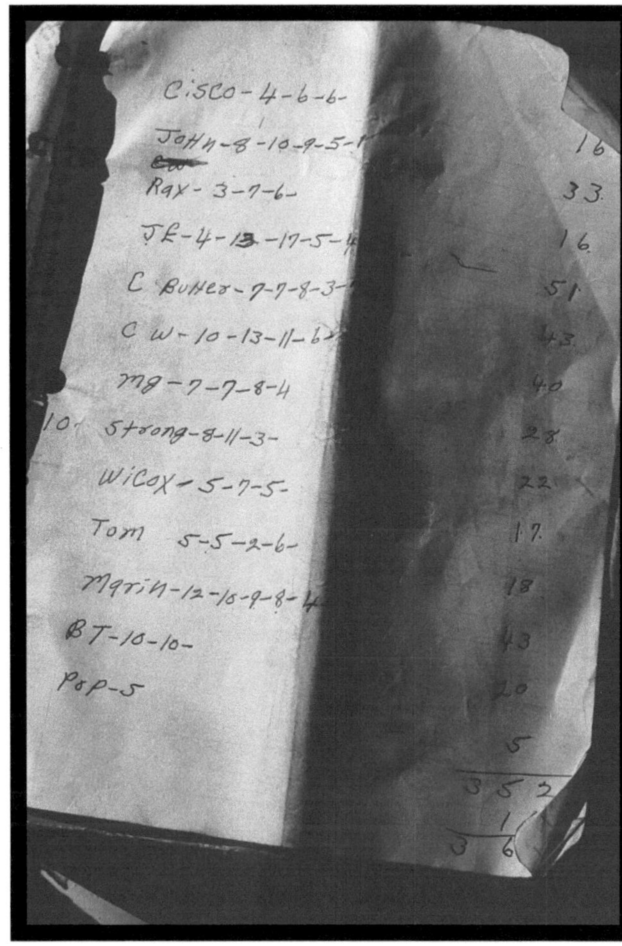

Cisco-4-6-6-
JoHn-8-10-9-5-
Rax-3-7-6-
C Butter-7-7-8-3-
C W-10-13-11-
mg-7-7-8-4
Strong-8-11-3-
Wicox-5-7-5-
Tom 5-5-2-6-
Marin-12-10-9-8-4
BT-10-10-
Pop-5

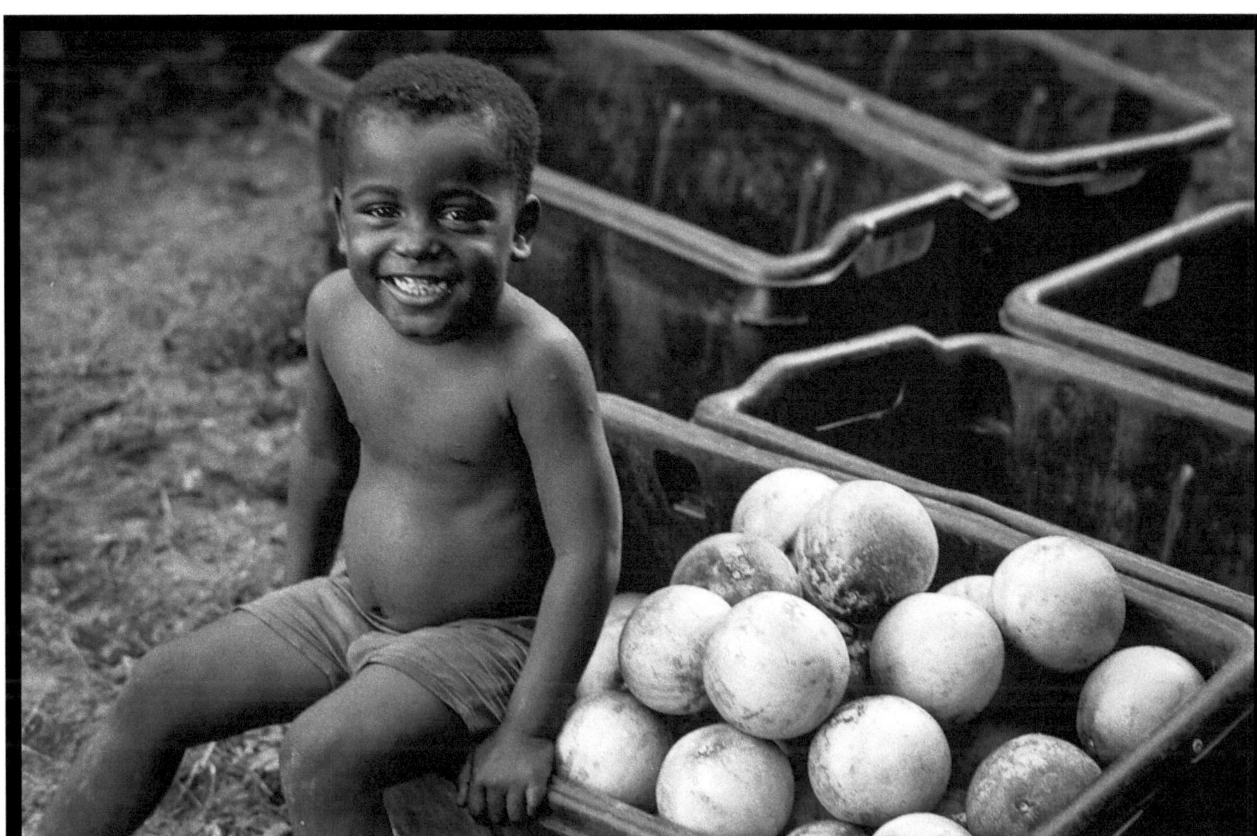

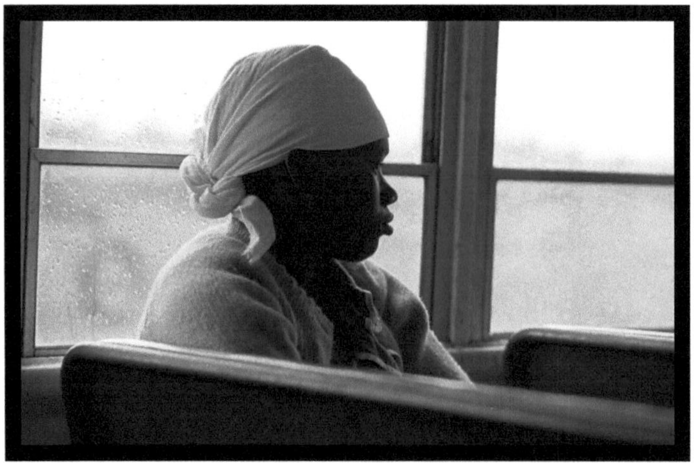

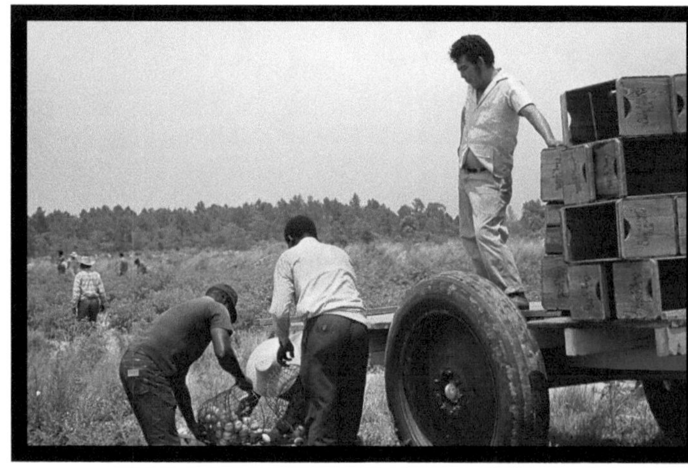

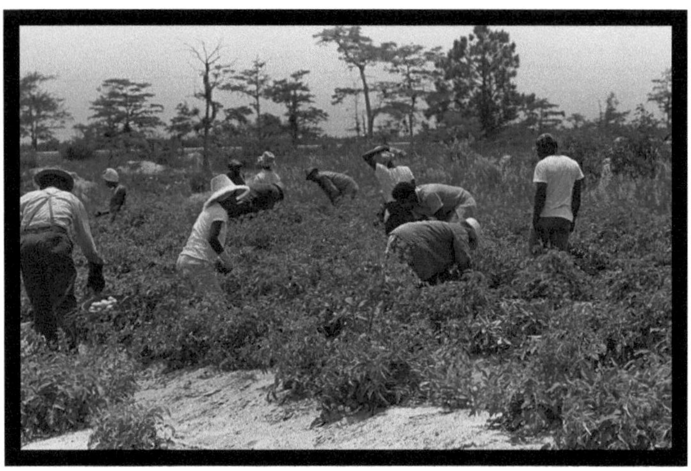

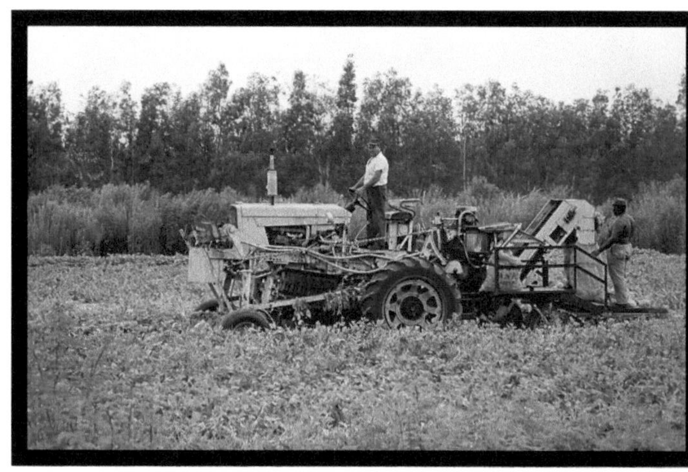

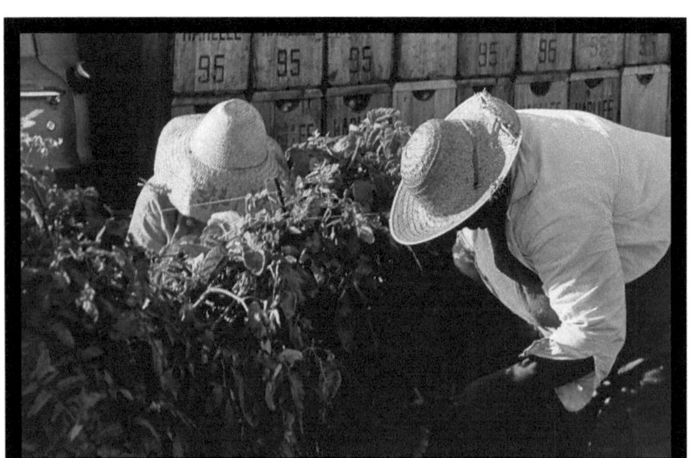

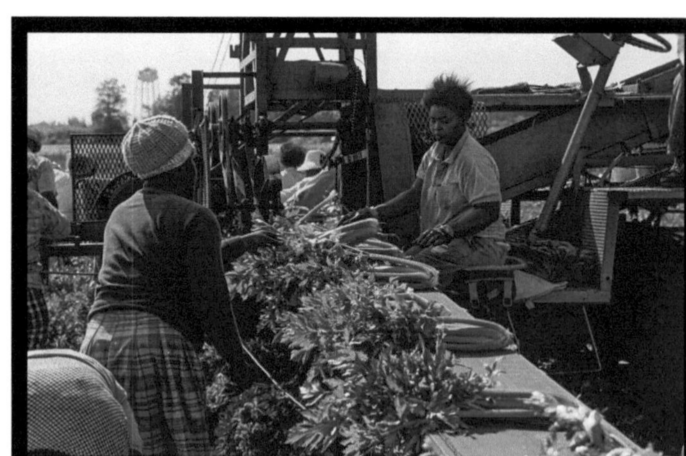

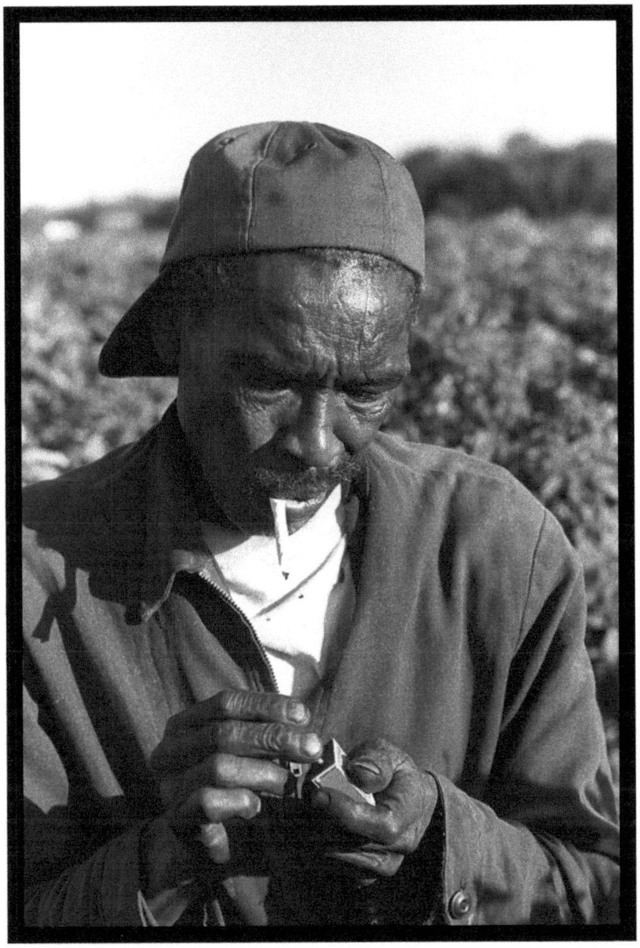

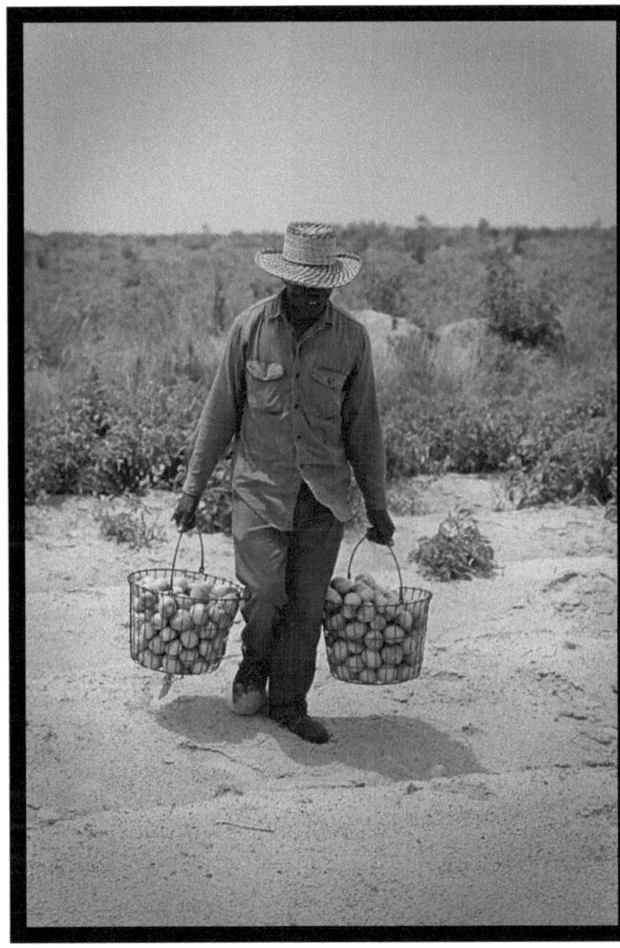

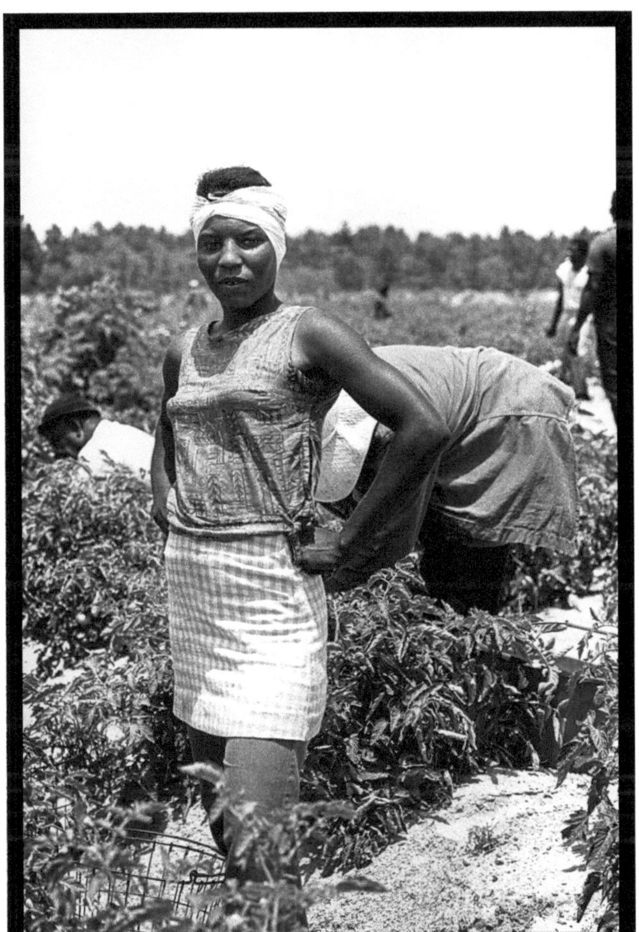

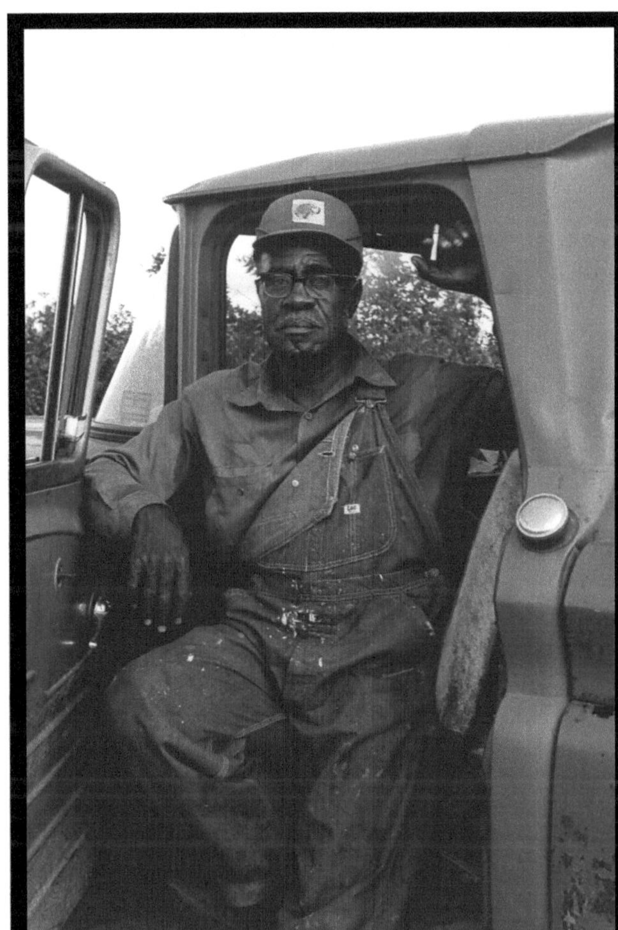

Tom Policano

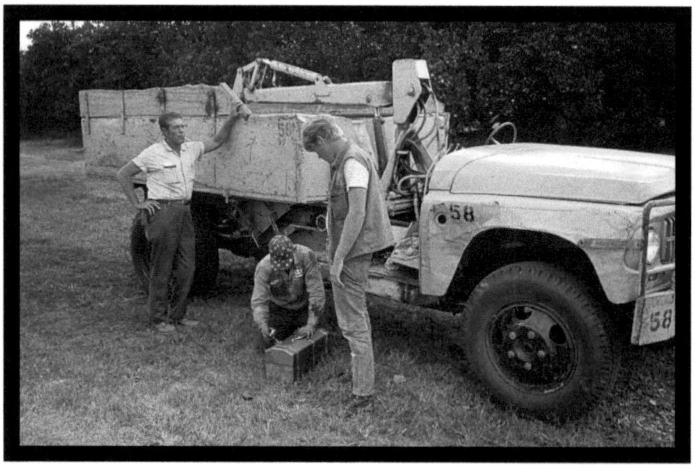
58

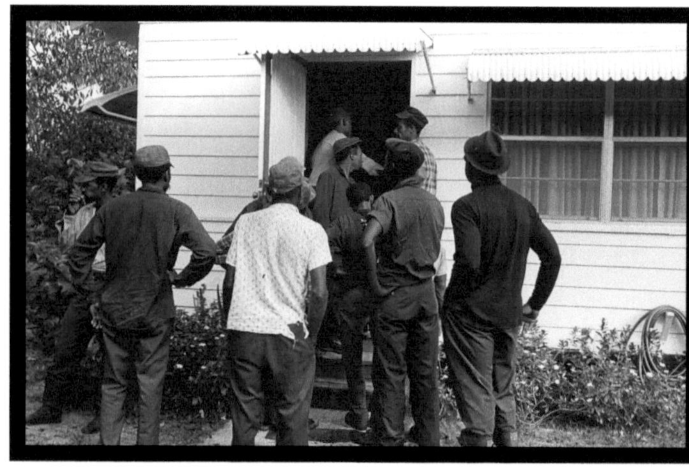

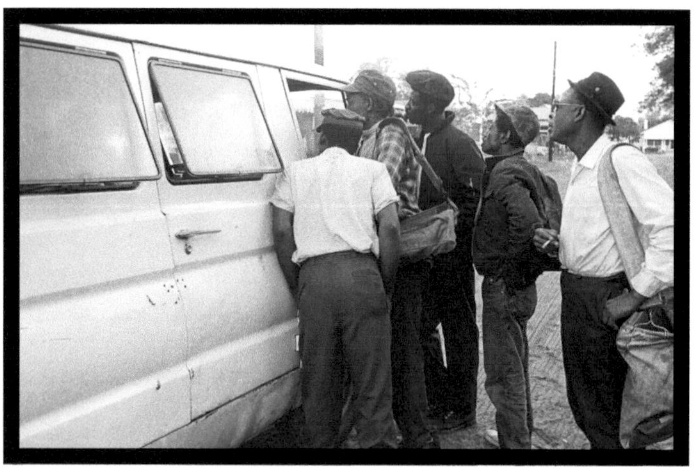

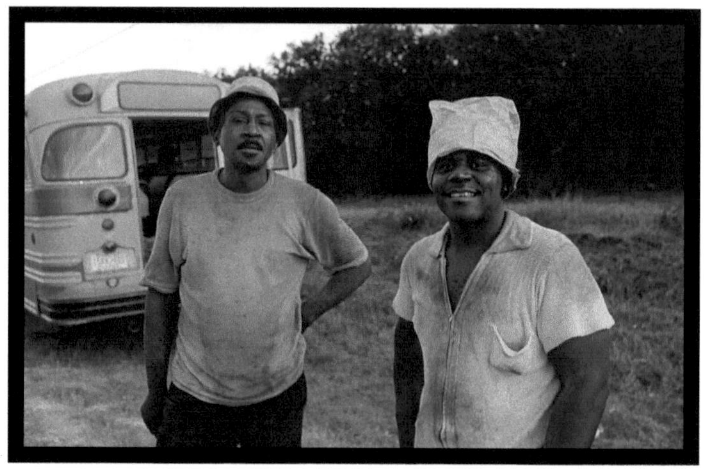

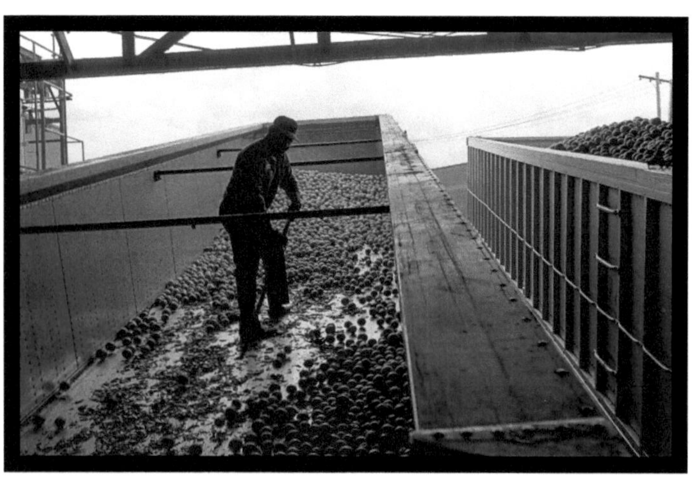

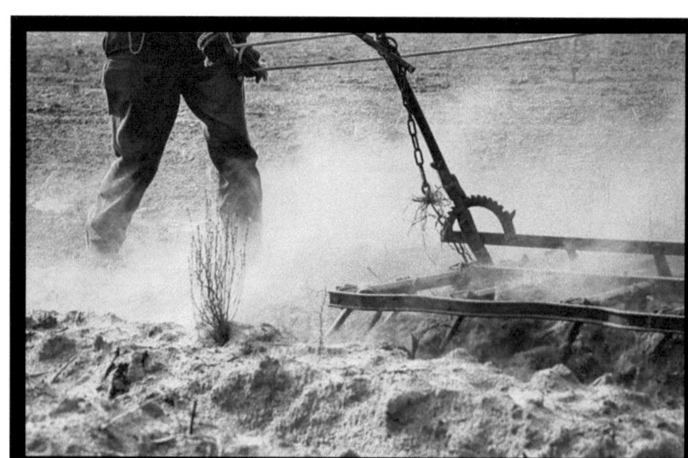

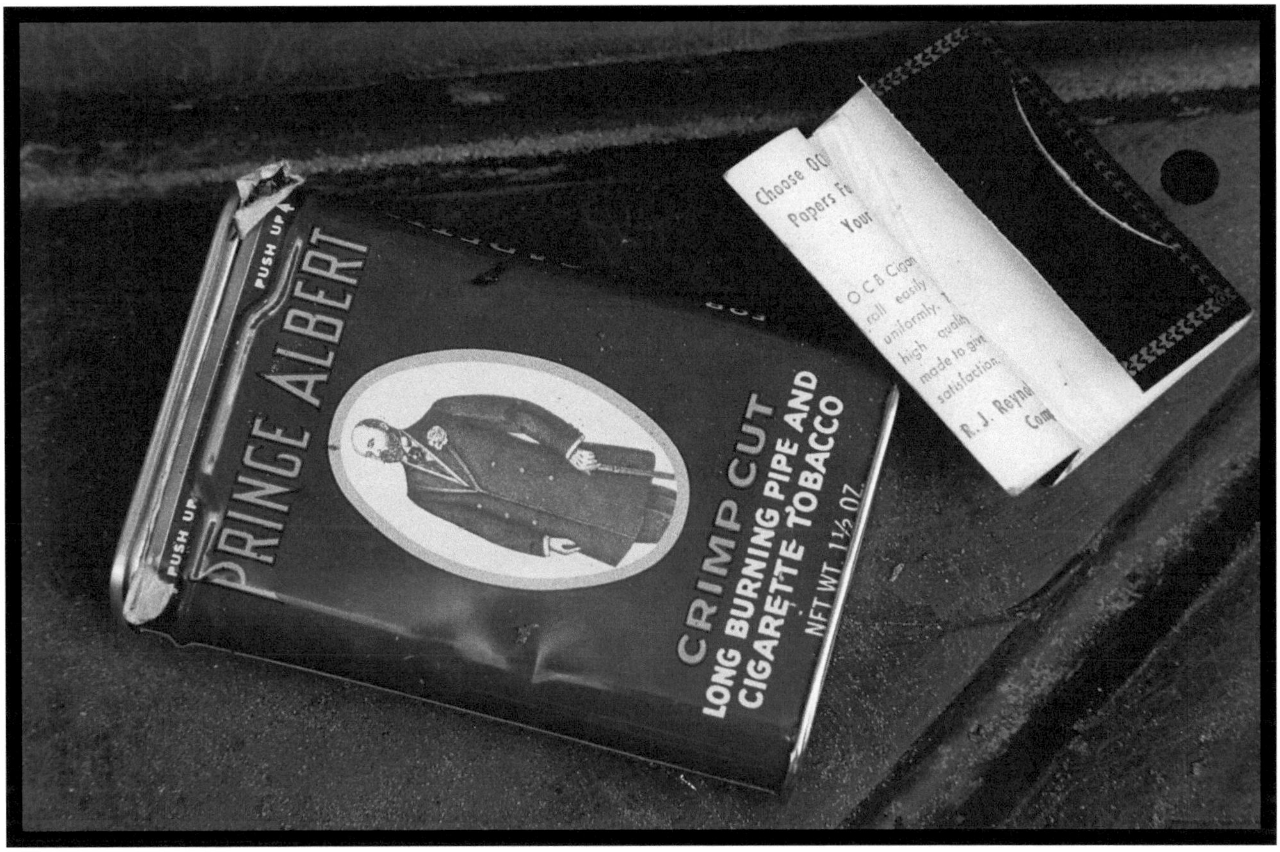
PUSH UP
PUSH UP
PRINCE ALBERT
CRIMP CUT
LONG BURNING PIPE AND
CIGARETTE TOBACCO

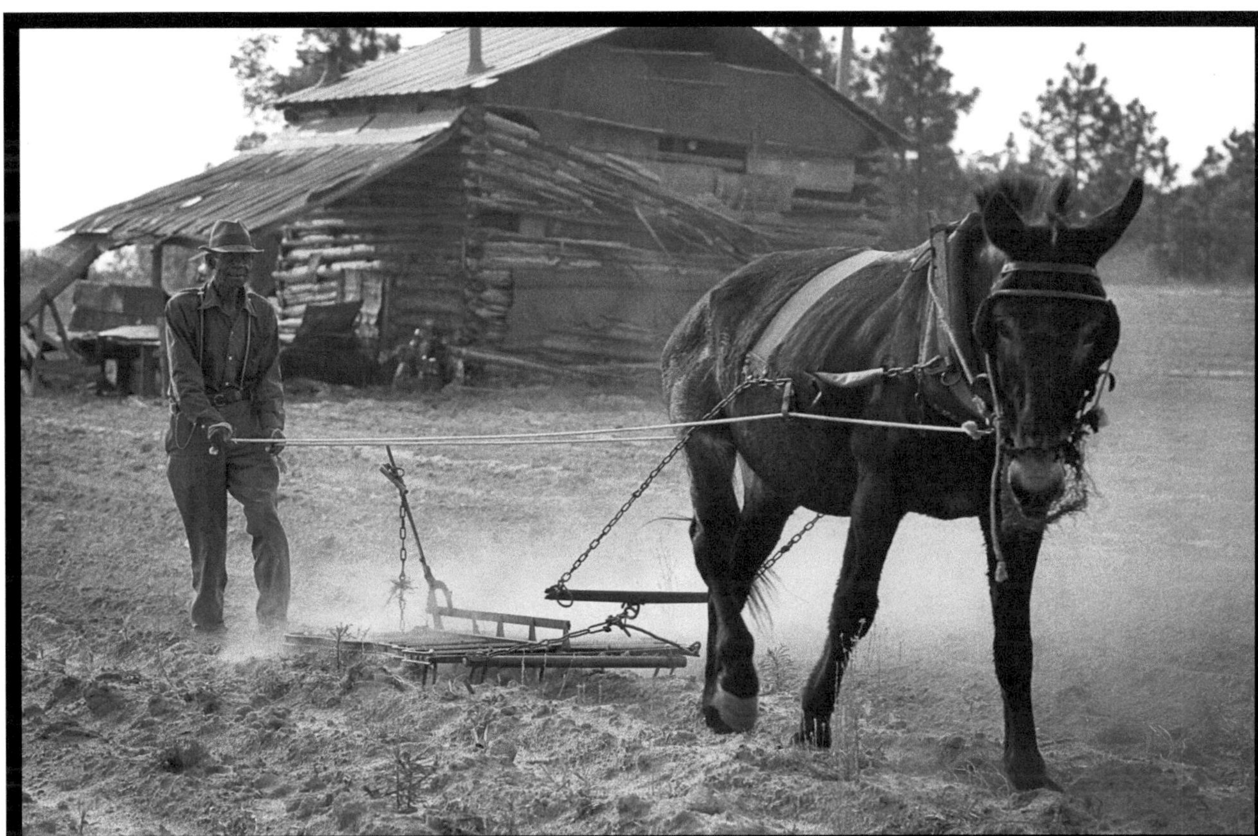

South — At Home

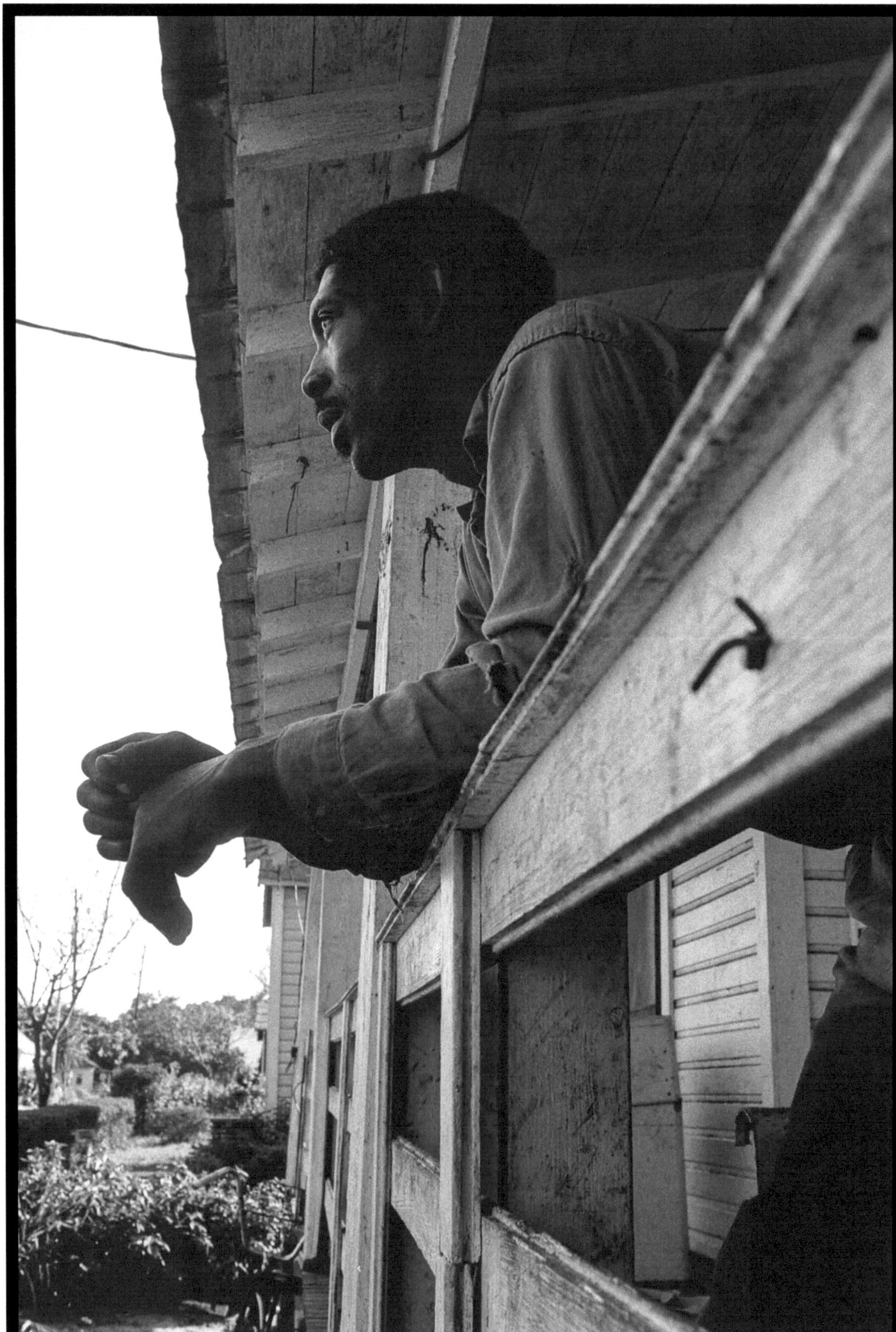

Slim Pickins

Johnny Wilcox

Ma habit has gotten me sort of down to this type of deals you know, cause I couldn't hold down a position of sponsibility drinkin liquor. There's nothin you can ever accomplish in life if you drink, I means excessively. So let's face it's this trapped circuit that I'm on is about the only thing that I can support ma habit on. Now if I ever decided to just stop drinkin period, then I really mean to get out of it. I've got to first stop drinkin. But it's hard to stop drinkin when the first thing you think about when you get up in the mornin: "Damn, I wisht I had a drink," an the las thing you think about before you got to bed: "Damn, I wisht I had another drink."

This is the thing that started when I was young. I was raised in the slums. First I was a farm boy. An ma father died when I was eleven years ole. An that left ma mother with a problem, had seven children, an she had never experenced buyin a week's groceries. She didn't ever know to go into town an buy a week's groceries. Never had to, see. An we'd ohways lived on the farms, raised most of our stuff. An daddy ohways done all the business an all the buyin. That left momma quite a problem. An she had no means of support for us, because ma father refused to be committed to a sanitarium. He died of tuberculosis at the age of thirty. An he was the father of seven livin children at the time.

We had two deaths, two children dead. Thas nine ahtogether. An ma baby sister next to me, she died at three months ole with pneumonia. I had another brother was stillborn. An he'd a been ohmos thirty years ole now. His name was Robert Lewis. I named im. I decorated his grave with goldenrod, if thas what you want to call it, I don't know, that yellow stuff, you know, flowers. I picked em, decorated his grave. Well, I named im because nobody else seemed to have time. Just a baby, you know. An after ma father died, we left. We were livin'n Guthrie, Kentucky. The las year ma father lived, we hadda move from the farm cause he wasn't able to raise crops. So we moved inta town.

An I don't know where'n the heck daddy got money, but he managed to get some money somewhere. He used to gamble, but he never smoke an never drank. But he got tuberculosis from exposure, man. It gets pretty cold in Kentucky, sittin aroun those ill-heated joints all night, two or three days at the time, losin his rest, not eatin proply, stuff like that. Thas what gived im tuberculosis. Then he felt his sponsibility to his famly was so great that he couldn't take the time off to go into the sanitarium. He had to stay there'n try.

Now things are gettin better, yeah, man. I think so. They're gettin a lot better. You take, ah, the fruit we're pickin now forty-five cents. Ah right now. I've picked worse fruit an that for thirty-five. Place removed, the day I've picked worse fruit an that at twenty-two cents for that matter. When I first came to Florida thas what they were payin. That was top price. If you got twenty-five cents for oranges, they were bad oranges, man. Bout eight or nine years ago that was. Thas all. An this stuff now, you gettin a good price, I think.

Now they's some guys still payin thirty-five cents, like this guy Kenny. He went to the dog track the other night, you know, an he was so drunk that this other guy had to drive the daggon truck back. The guy parked over on Scott Street an tole Kenny don't move it til he sobered up some. But Kenny got behind the damn wheel an backed out an then nobody seen im since. So I figured he might be in jail someplace

I'm gonna pick oranges right on for a while. I really don't know what I'm gonna do after. I'm not zeroed in on anything definite. I'm disoriented. It's like if I was goin around that damn lake, I wouldn't know which way to go, which direction, let's say. If I had to meet a point directly opposite from where I started, I'd be undecided which was the best route to travel aroun that damn lake. Thas the way it is. There's nothin but fish in that lake. Rash an them fish all the time. They catch some nice ones sometimes. I've never tried it.

But I'm disoriented because I'm not motivated strongly enough. Maybe if my situation was different, you know, if I had somebody insisted that I do straighten out, maybe would try to help me straighten out little bit, I might go for it, you know, get outta the stream.

Johnny Junior was motivation at one time, but anymore he's gotten to be sort of common. Sort of like all the rest of the kids aroun, you know. He's two months, two months, man. He's common. See, at first he was a little somethin, little somethin special. Ain much special bout him now, cept he keeps me awake half the damn night screamin. I didn't have a drink, I couldn't sleep. Not in that house, man. How could I sleep in that, man? Only way I can sleep is when evrybody else is already gone to sleep. Or pass out, yeah. Then I can sleep.

I went to pick tomatoes, you know. Thas how I met Claudice. Her'n her mother was workin together. An she was doin all the work. Mother was standin up an lookin at the god damned sky. An she ast me if I'd lug for em. So I started luggin for em. Shit. I said, "What the hell. " I lugged their tomatoes for em, both of em. An I picked for em. We'd split the money three ways. I caught here down'n the damn grove there, an we kissed a couple times. An I said, "Shit, mighs'well."

We married. I don't know why. I musta been drunk or I wouldn'ta ast nobody to marry me then. I was drinkin an I ast her to marry me. Well, I ast. I think a lot of her now cause she does anythin in the world for me. But sure as hell I didn't care no more for her then. I jus doin somethin. "Let's marry. Let's get married." I went ahead, up'n married her. We had a little small weddin over the justice o peace n Florida. We been married three years, ohmos three years. Humph.

One of these days I'm gonna go up on a season an won't come back. I was tempted to do that this year, but the baby was too young, evrythin, but I started to do that this year. I had ohmos got me set up. I was gonna leave the camp an get m'self a job up there, but I don't want the sponsibility. You got a steady job, forty hours a week, you gotta be there, man. Can't be bullshittin about. On this season shit, camp shit, I don't feel like workin, I say, "The hell with it." Nobody's to bitch. I might go back up. I don't know. I'm not sure what I'm gonna do yet. I'm ohways unsure what the next move's gonna be. Never sure.

If I can have what I wanted right now, I'd like to own me bout acre of land somewhere in Florida jus far enough away from my damn in-laws that they couldn't walk there or couldn't ride in an hour or so. Now this I would enjoys to get the hell away from them folk. An I could take that acres man, an fix me some trailer houses on it. I could fit about, with an acre of land, I could fit about eighteen of them buses. What I'm doin, I would take em an remodel em, you know, the outside, partition em up, seal em off, make good trailer houses for season people. You know, people goin, comin, you know, season. I'd like to do it somewhere on the east coast or somethin, doin one at a time, where there's somethin doin, you know, I wanna retire.

Well, Ninous wants me to try to get a steady crew, you know six or seven guys that'll stick with im, you know, an not look for their money evry night, but he'll have money, you know, to advance em, he said, when they need a little somethin durin the week. I tried to explain to him that the situation in Lakeland is not like it is in Palmetto, see. In Palmetto, his mother runs a restaurant there. Ah right, an she sells wine an beer. Thas all a fruit picker needs, really. I mean, when he runs outta money he needs somethin a eat, somethin a drink. Thas his biggest problem, you see, when he wants money before pay day. Ah right. In Palmetto the guys go to his mother's shop there, an she lets em have it on credit. An Ninous pays em off, usually he pays em off right there in the shop, where she can collect it, you know, before they get away. An Smitty does the same thing. But the guys here, they have a diffrent situation, man. Have to buy their meals from the restaurant down there. Most of em don't even have a place to cook. Thems do have a place to cook sure as hell have to buy somethin, you see. An Miss Pearl serves meals, fixes sandwiches for lunch an all that stuff. An she lets em have their wine, so much a day, you know. An it's quite a diffrent situation. An I tried to explain that to him. An he's undecided about Granger, you know. He don't know how Granger feels. An he thinks that Granger is fixin to quit, which he might, you know.

Claudice insists on goin to work with me tomorrow, cause if I'd a worked, you know, I coulda picked ten more anway, ten more, you know, if I'd a gotten wih't. I was sort of disgusted. I hate to wind up a grove anyway. It's sort of disillusionin to start with, you know. You think you got four or five good trees in front of you, an someone comes outta nowhere an cuts right across in front of you. Then to top evrythin off, first off this mornin went out there, someone took m'ladder. Okay, took ma ladder an took most of ma boxes. I had quite a few boxes there. So Stacy was just hoggin up the whole damn grove. Quite amazin how that man works. An Stacy, he works like that ah the time, man even when he's runnin the crew. I've seen'm load fruit an pick at the same damn, I mean, load fifty, sixty boxes an load three or four trailers. He seems to never get tired. He does, you know, but he never shows it. Then after gettin goofed up with ma ladder, Ninous got me a ladder was about two feet too short to top the trees. I butchered up a couple trees, climbed about like a god damned fool. That was discourag'n as hell. Then finely Granger found a long ladder over there, stood it up at the tree, an Ninous come by an tell me there's a ladder. I go down an there's no bottom rung. So I took it back up there an beat the rung out of one of them short ladders. So I put a rung in it. Tomorrow someone probly have it. But I walked that damn thing over here an I take it back, cause I put a lotta labor into gettin it so it'd work. An it's big enough to top any of the trees. An I like that.

Claudice, she went down to Salvation Army thing two weeks back when they was ceptin applications for Christmas, a needy family deal, see. She got a certificate to get her some meat or somethin, an then she got another certificate to go down an pick up some toys at the Salvation Army, you know. An she did that. She got some pretty nice toys this time. Las year she went through the same damn regime an come back with a bunch of broken-up junk. They'd gotten all the damn toys when she got there. Well, she lost two days' work. We coulda made fifty bucks. An she's messin aroun there with five or ten bucks of toys, damn an she couldn't even buy a turkey for six bucks.

Roosevelt was tell'n this joke, you know, in the grove. Said he'n this guy had an argument, an this guy wanted to know "Why do you talk like that?" He said, "I put ma hands on ma hip, an I said, 'Sometimes I talk like ma mother,'" an then he grabbed im in the collar an said, "Sometimes I talk like my god damn father." Boy, that Roosevelt's someth'n else. Roosevelt has personality. He might be a sissy, but he's not the kind you deplore, man.

Man, I've had to do some ugly things in ma life to survive, I meant I guess nature, I guess God works out for people, because I can remember times when I didn't know what the hell I was gonna do next, but somethin ohways came up. It ohways worked out. I'm grey-headed right now from worryin an worryin an worrying an I worry about things an there's no need at all to worry. So I finely come to the conclusion, did that today, gonna stop worryin bout other people. If I owe somebody, let them worry about it. Damn. Why should both of us worry about it?

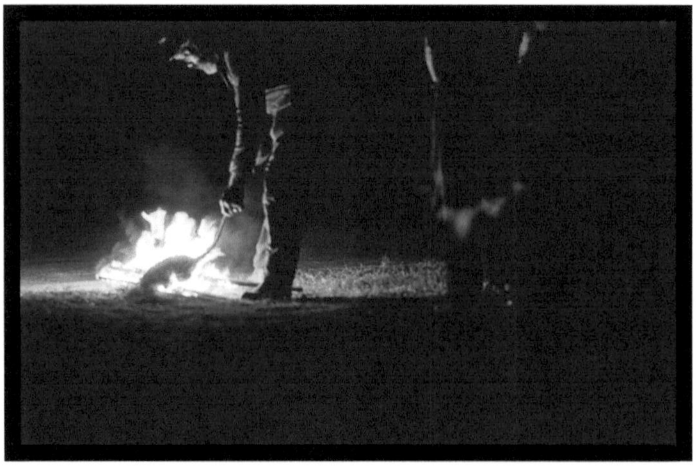
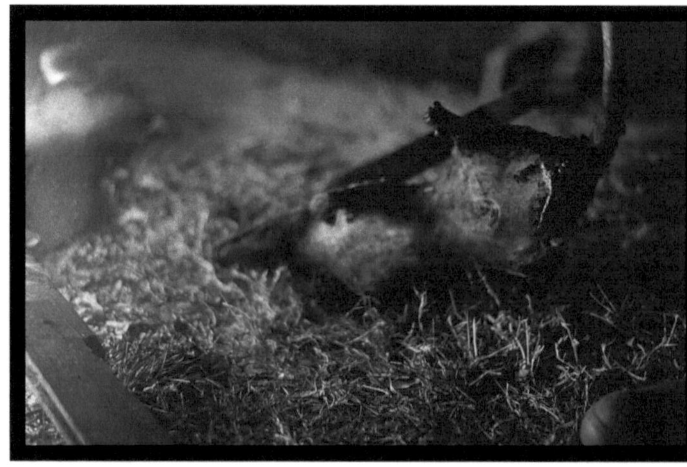
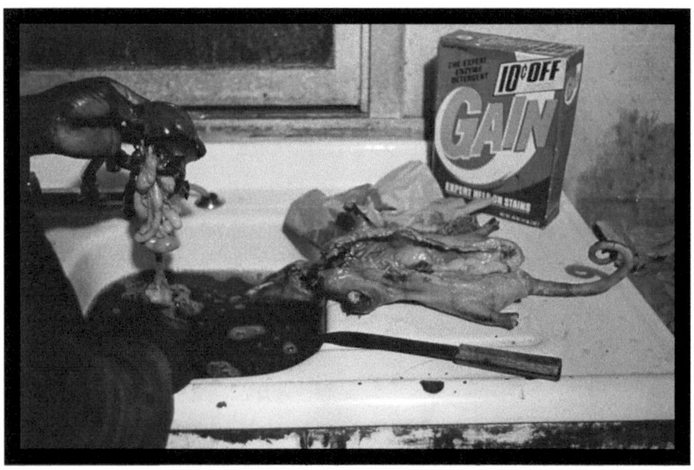
10¢OFF
GAIN
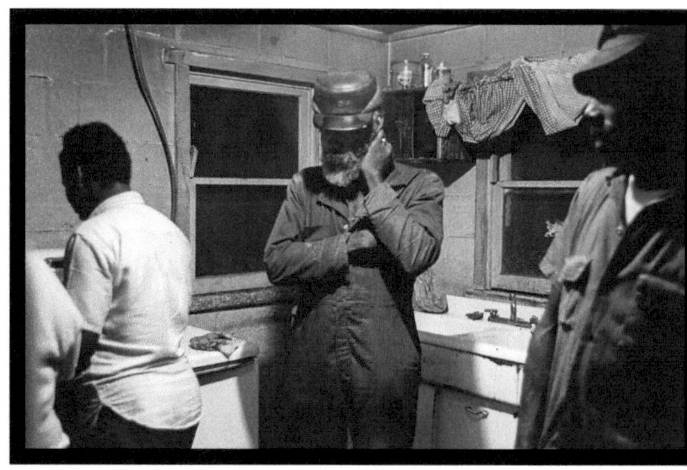
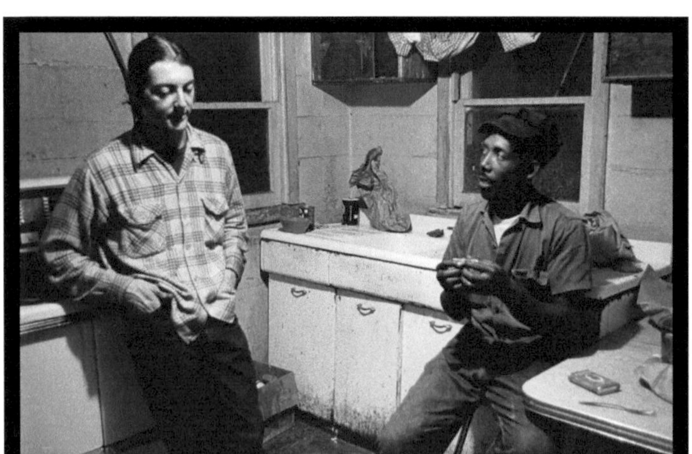
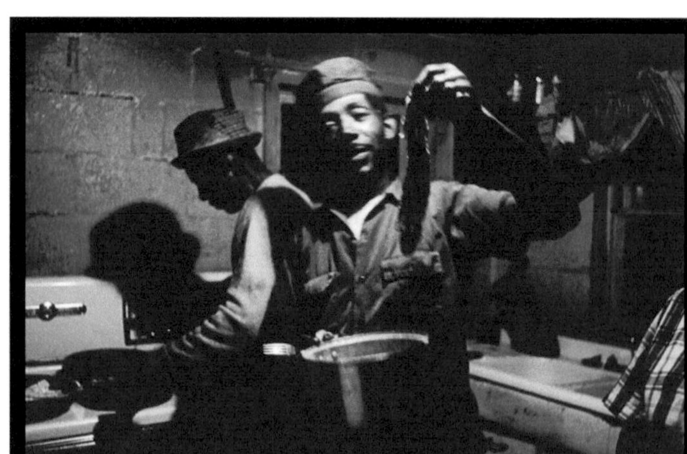

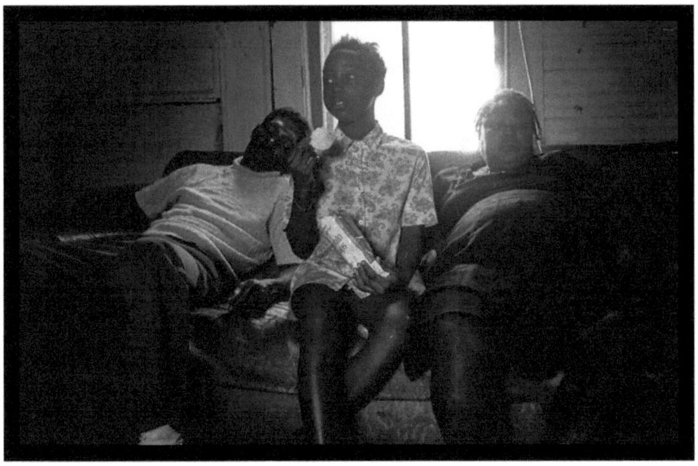

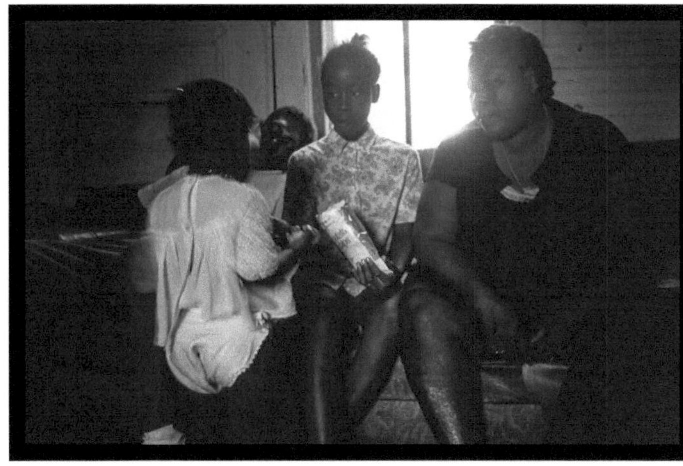

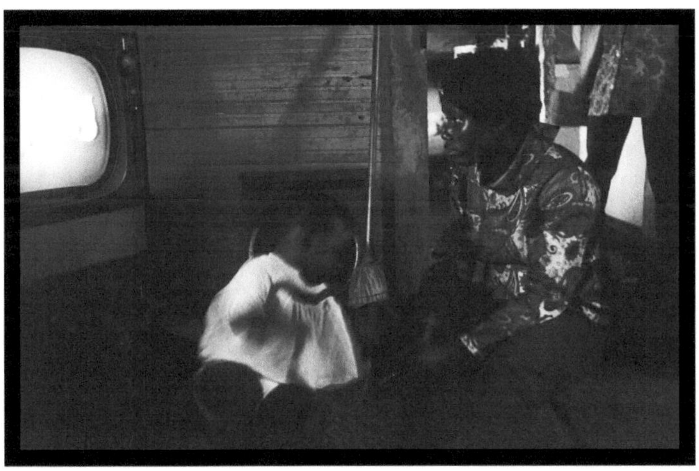

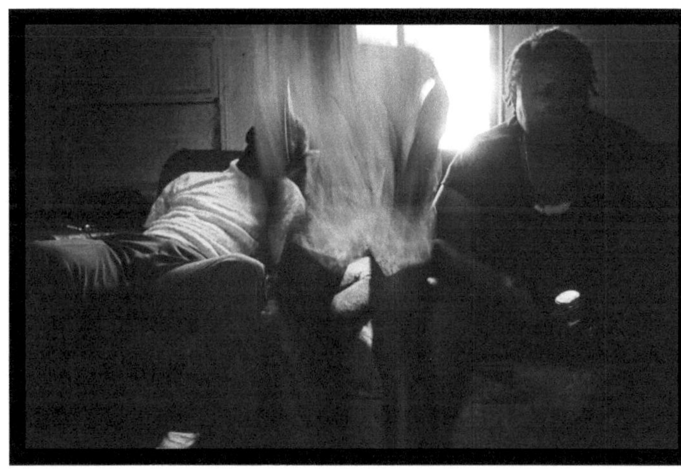

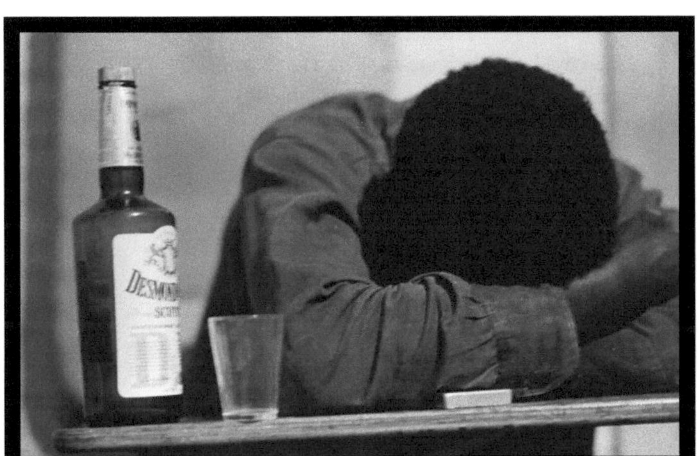

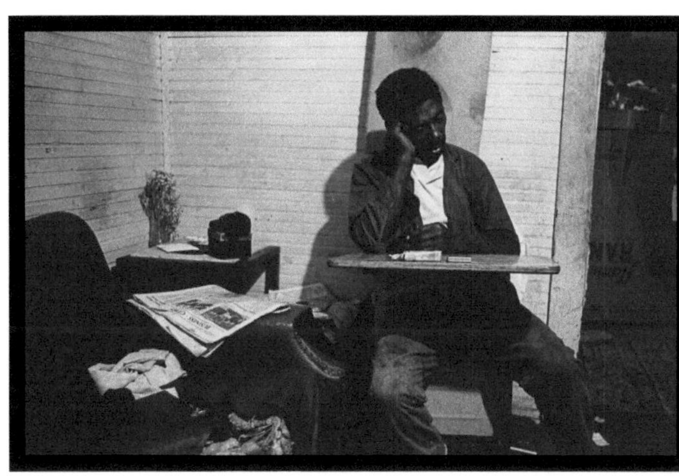

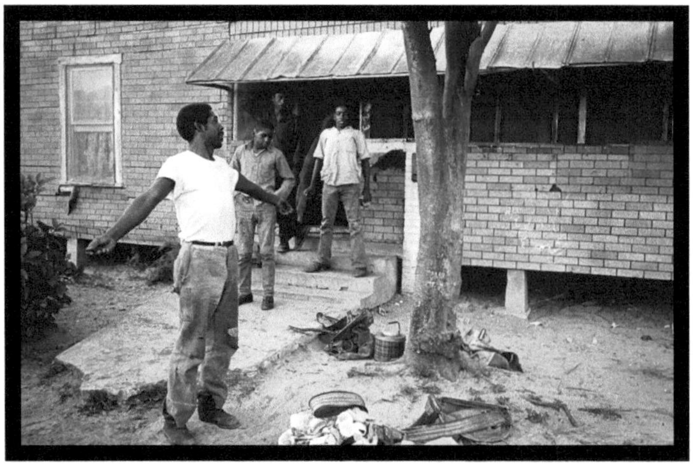

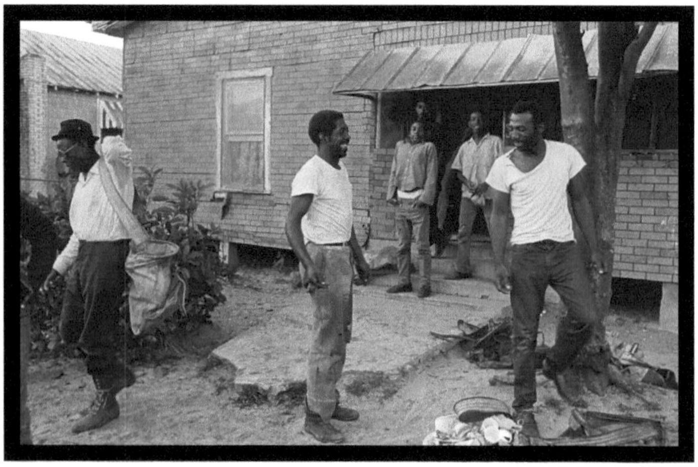

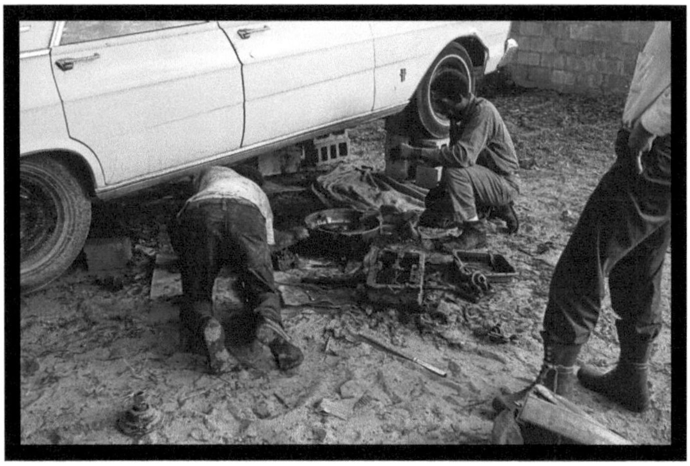

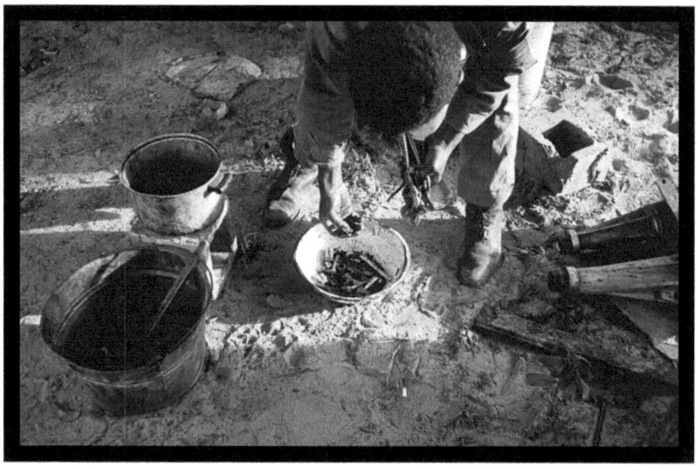

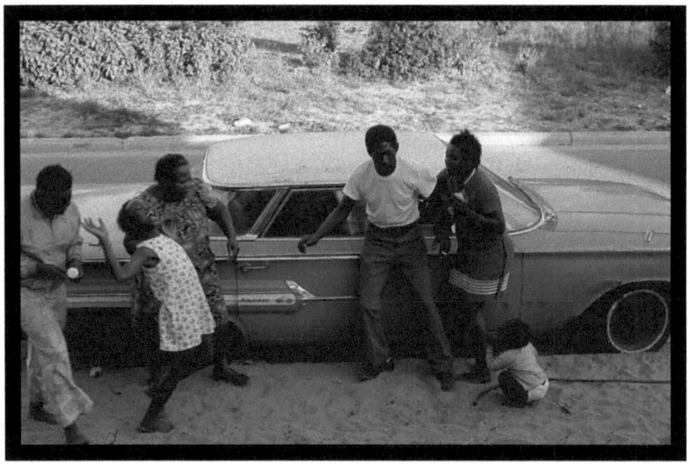

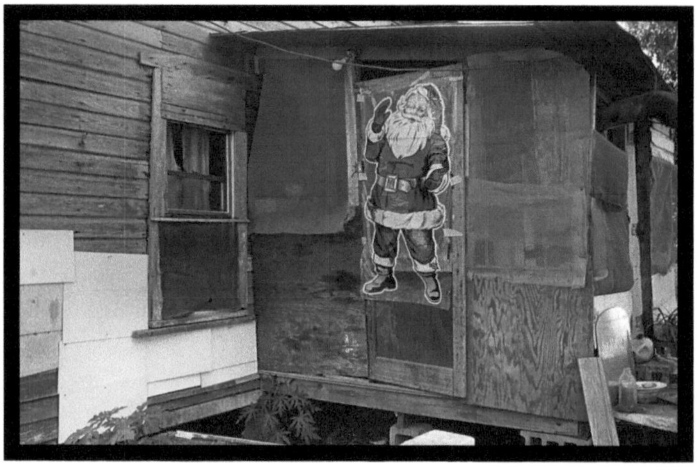

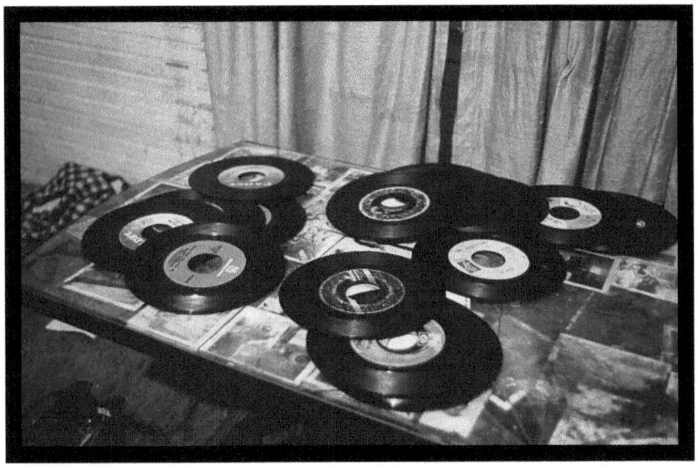

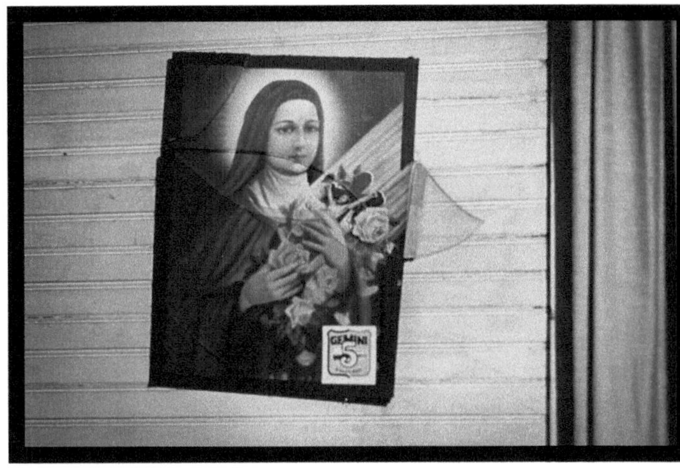

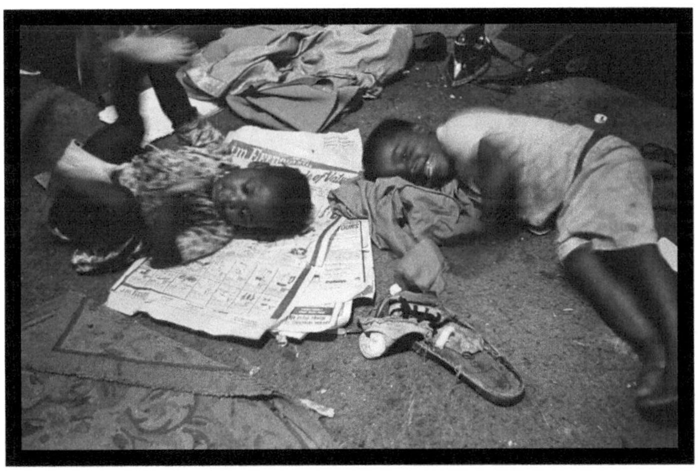

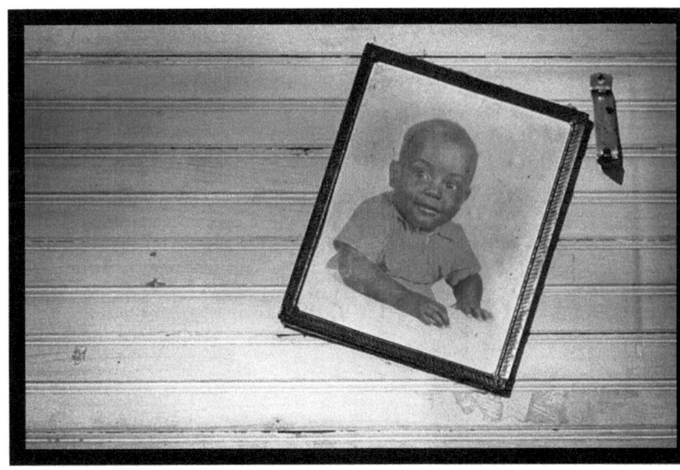

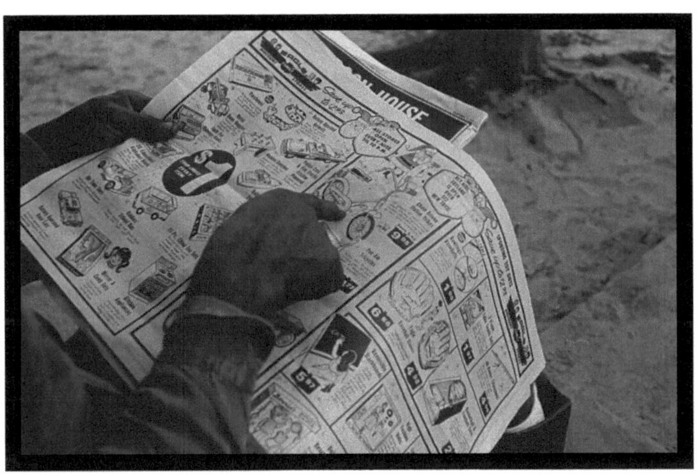

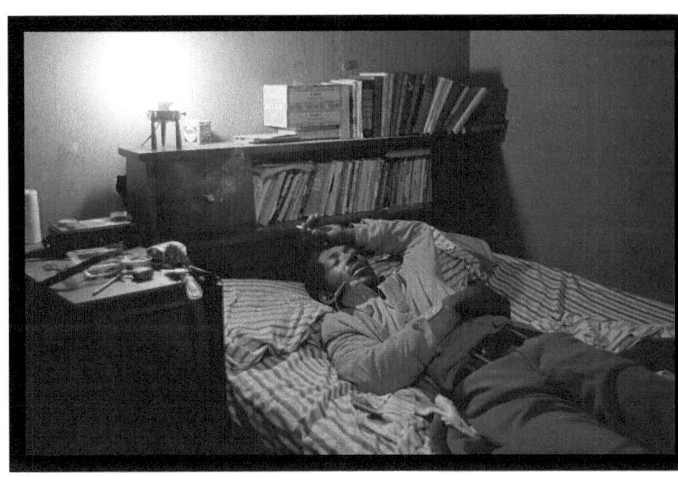

Tom Policano

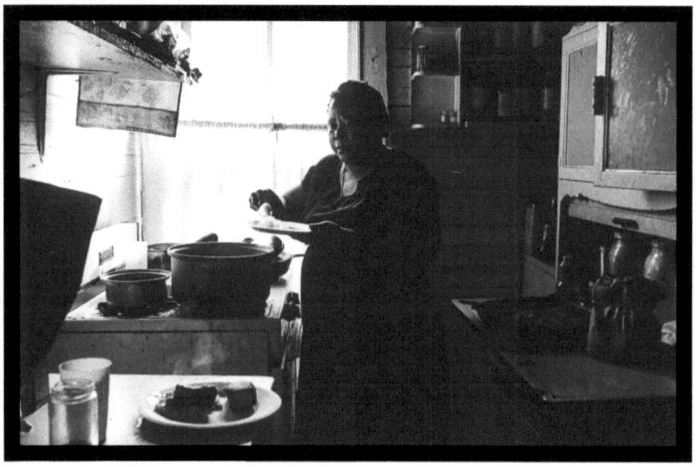

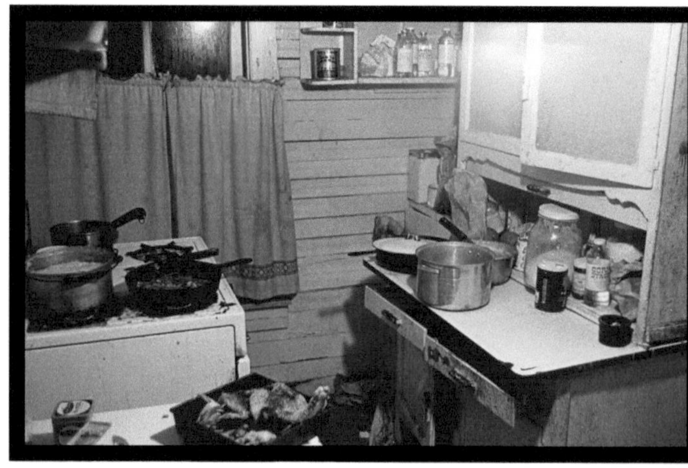

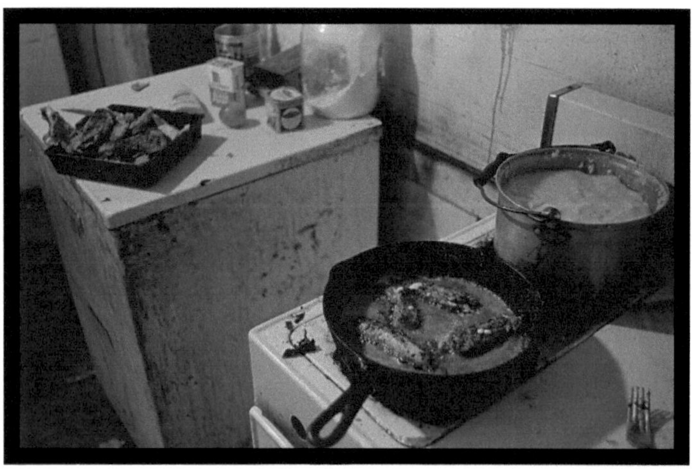

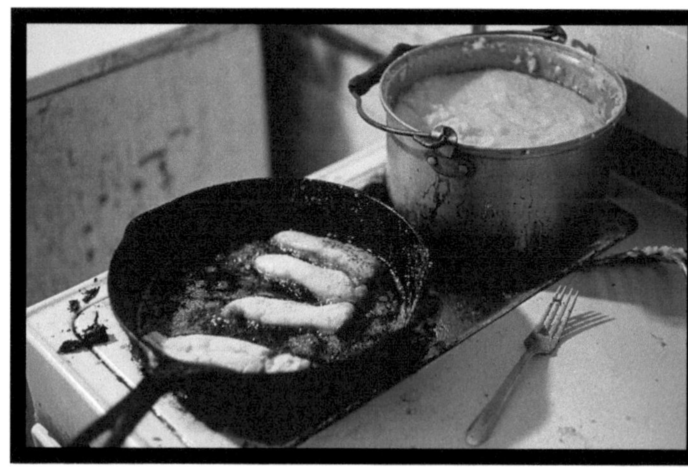

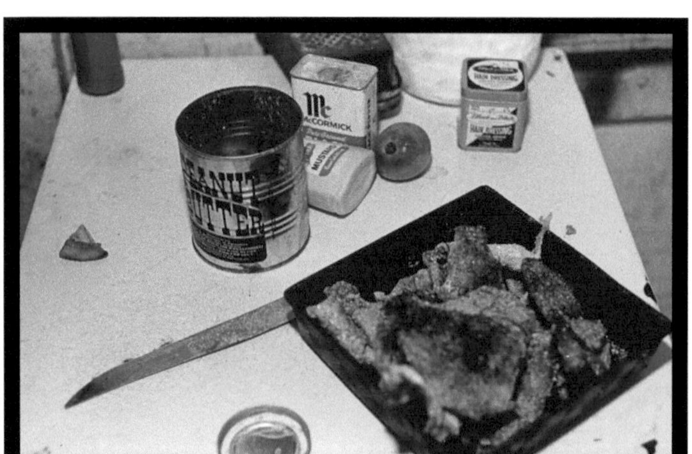

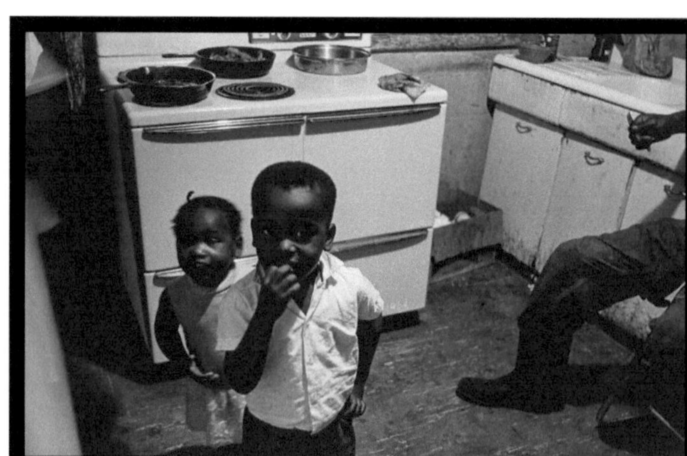

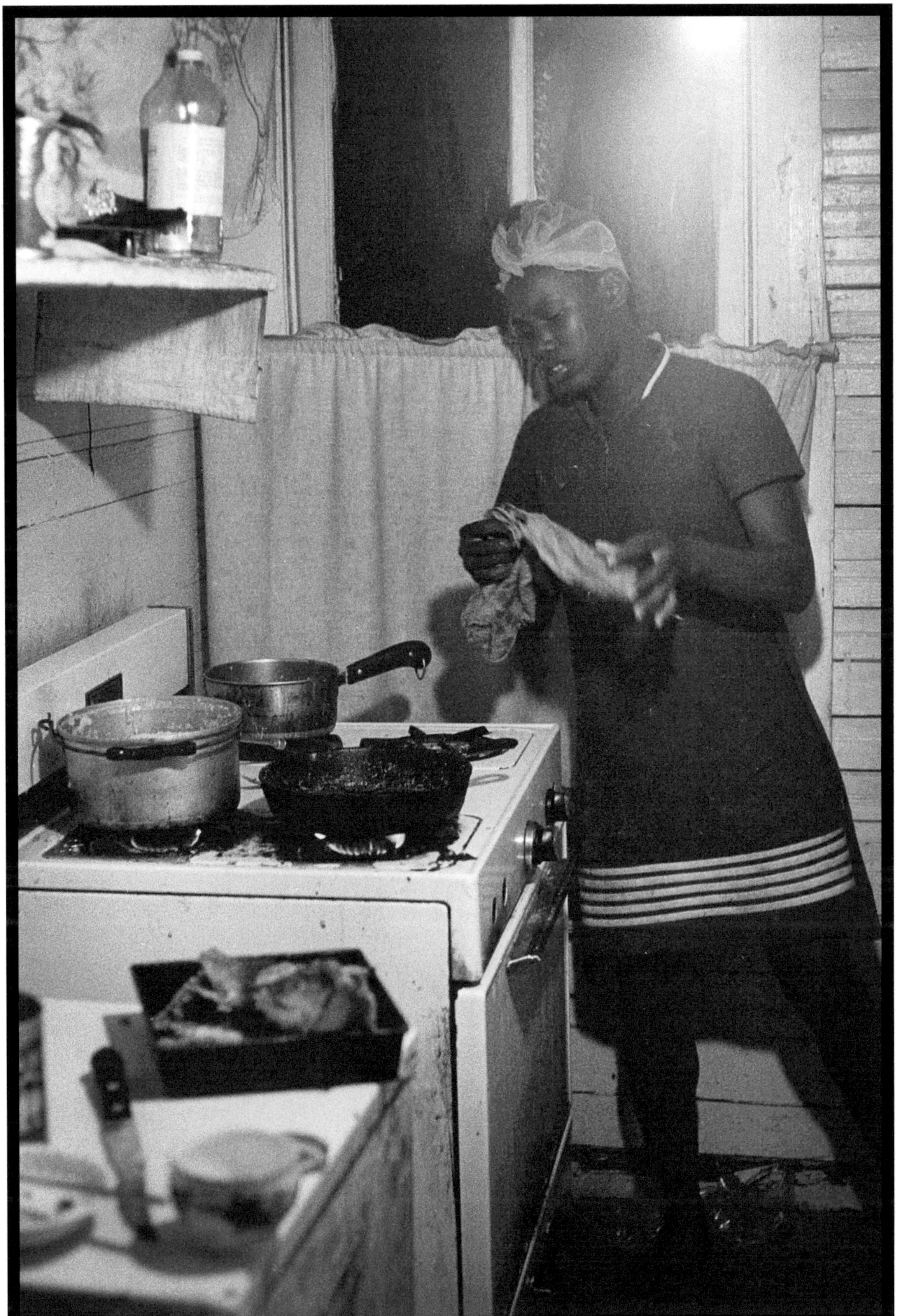

Tom Policano

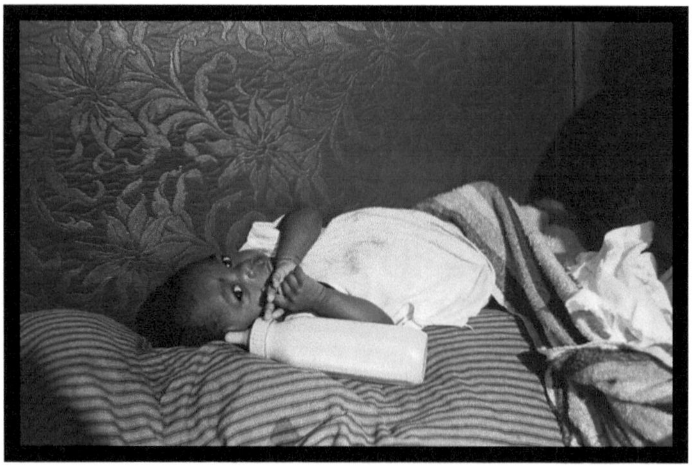

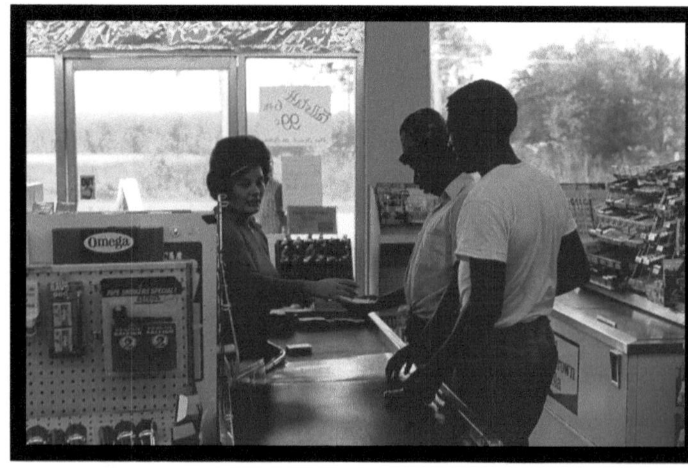
Omega

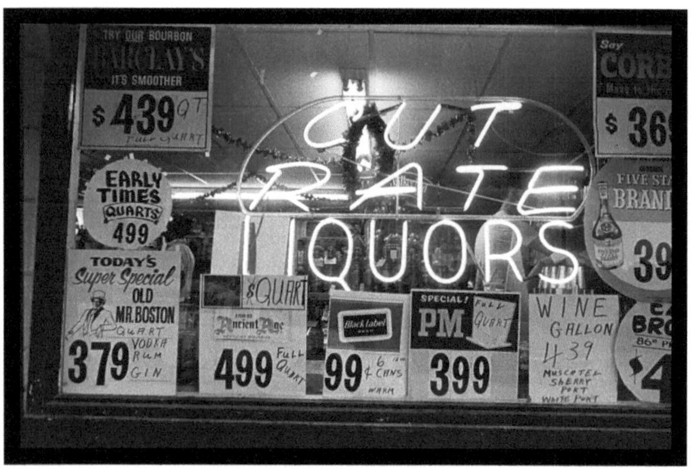
TRY OUR BOURBON
BARCLAY'S
IT'S SMOOTHER
$439 QT
FULL QUART
CUT
RATE
LIQUORS
EARLY TIMES
QUARTS
499
TODAY'S
Super Special
OLD
MR. BOSTON
QUART
VODKA
RUM
GIN
379
$ QUART
Ancient Age
499
FULL QUART
Black Label
99
6 & CANS
WARM
SPECIAL!
PM
FULL QUART
399
WINE
GALLON
439
MUSCATEL
SHERRY
PORT
WHITE PORT

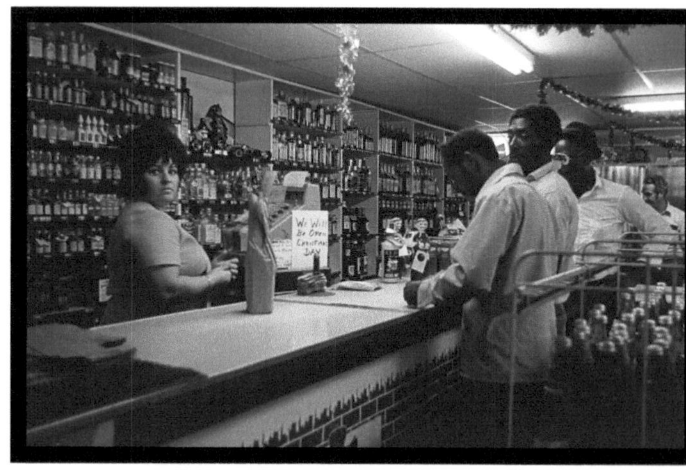

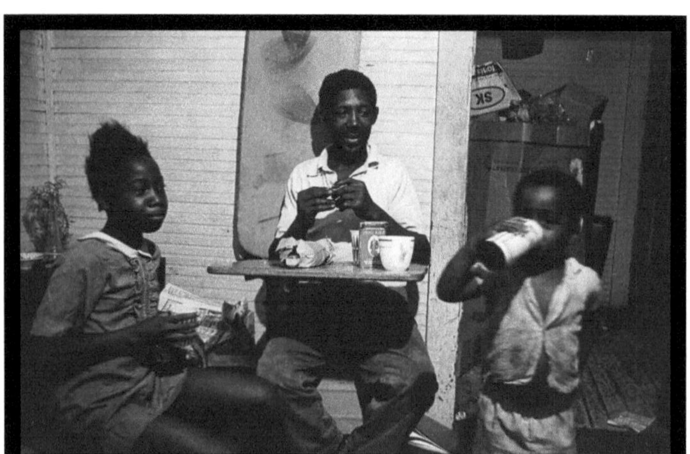

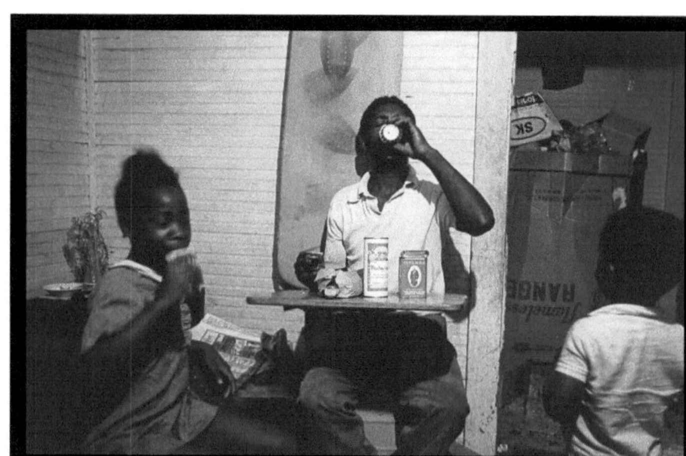

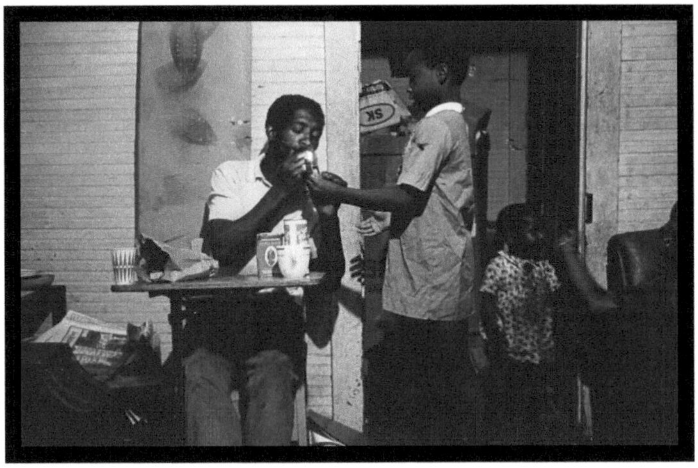

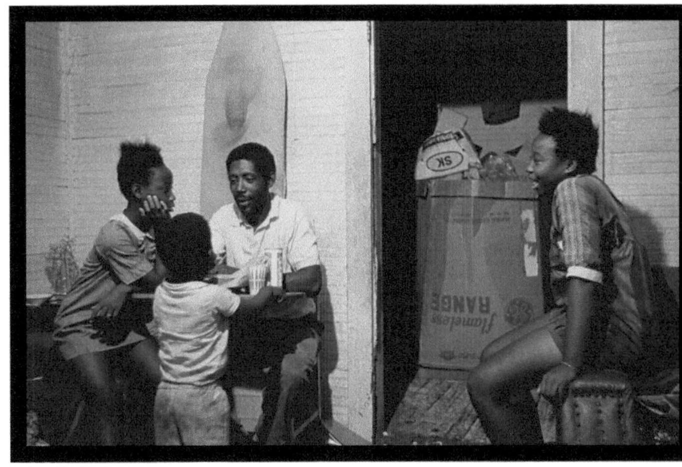

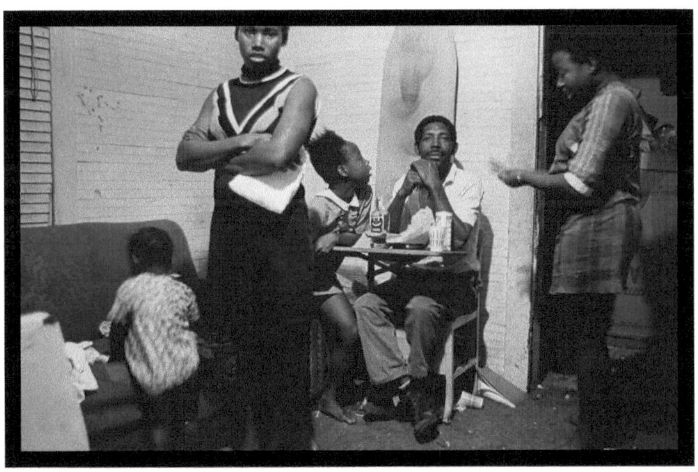

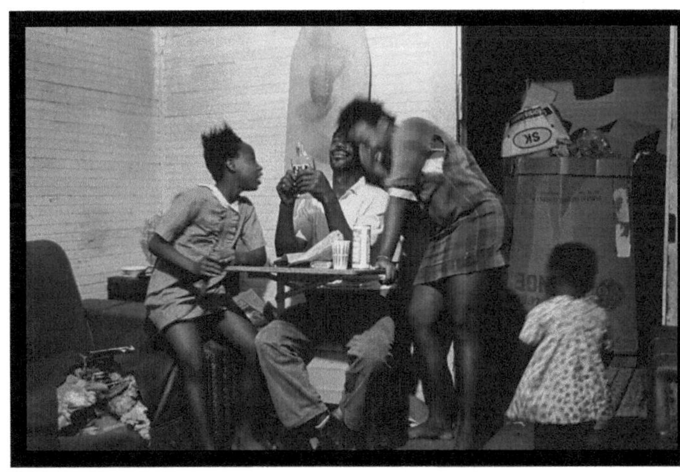

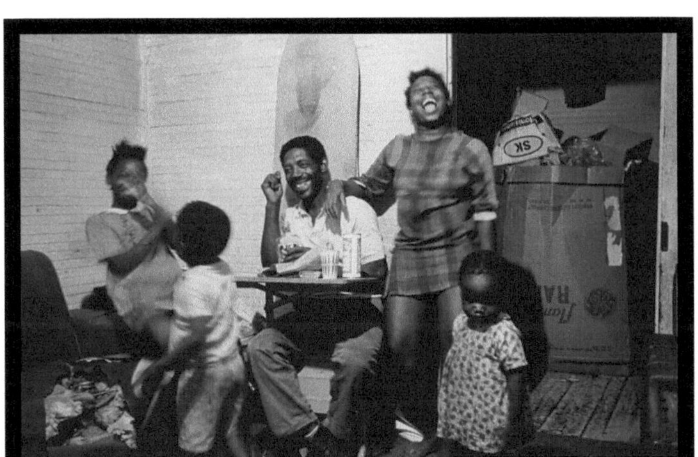

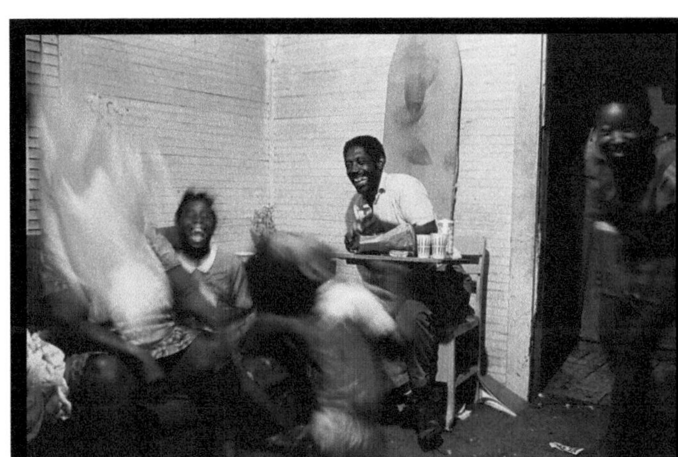

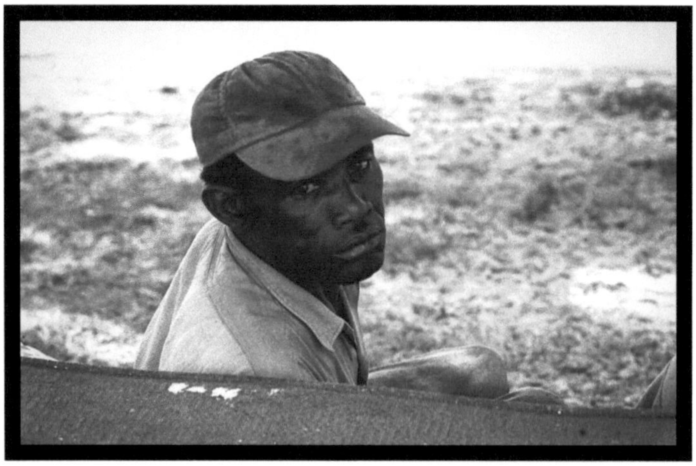

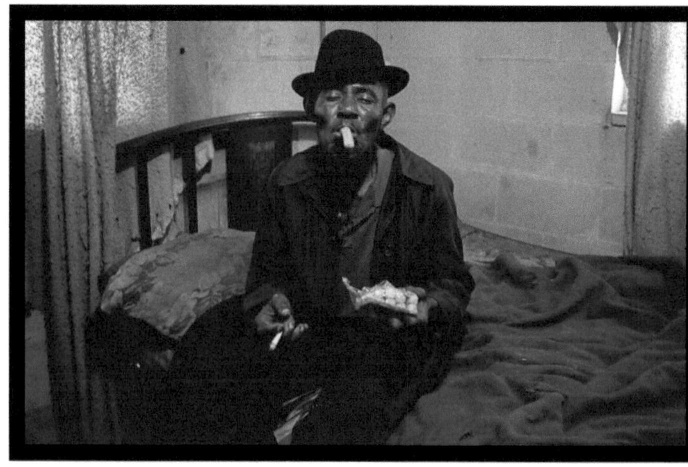

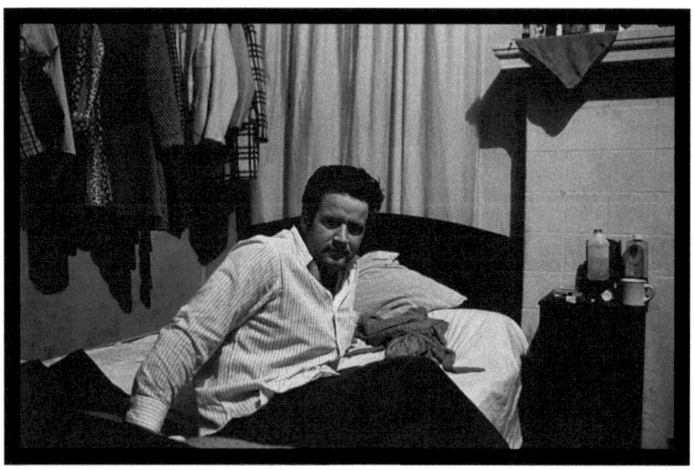

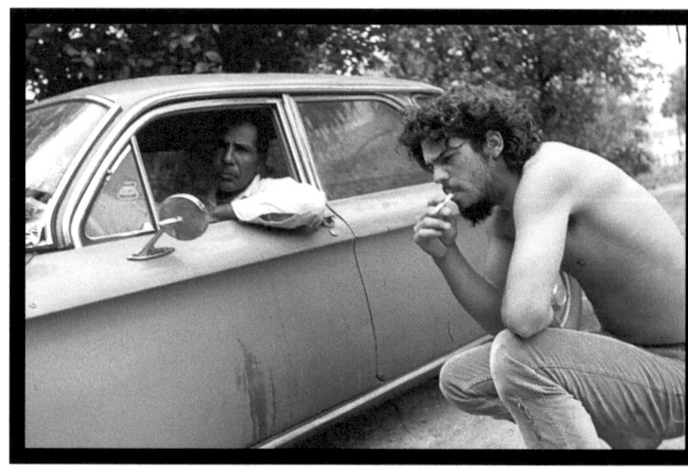

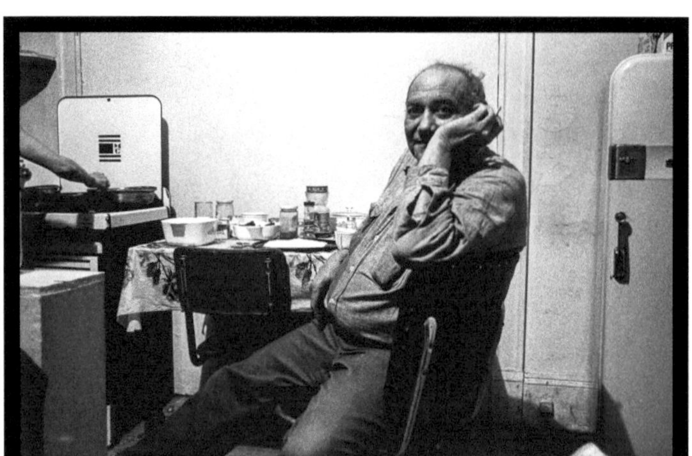

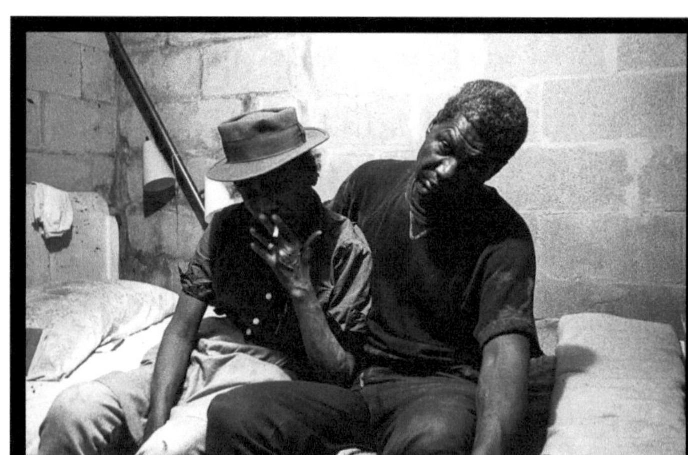

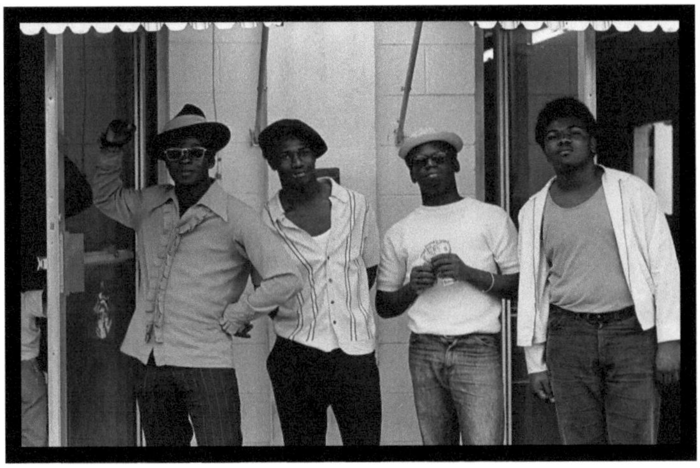

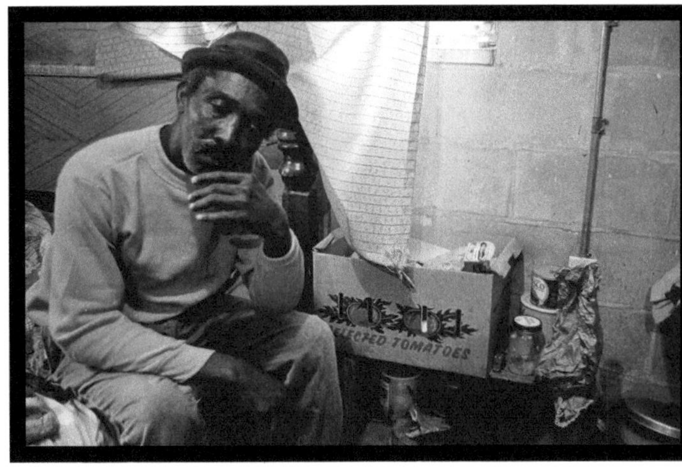

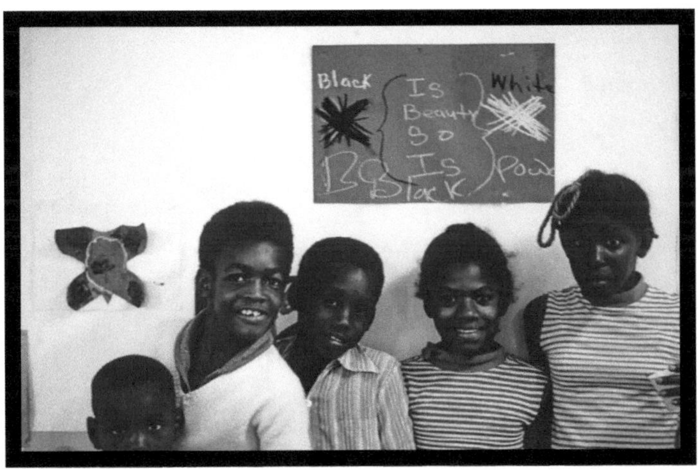
Black
Is
Beauty
So
Is
Black

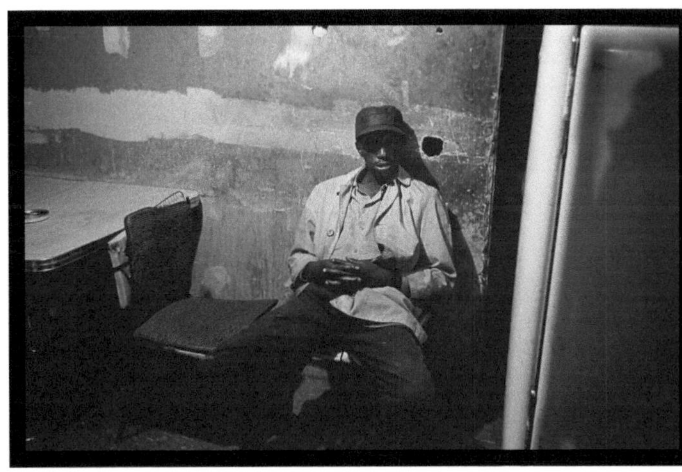

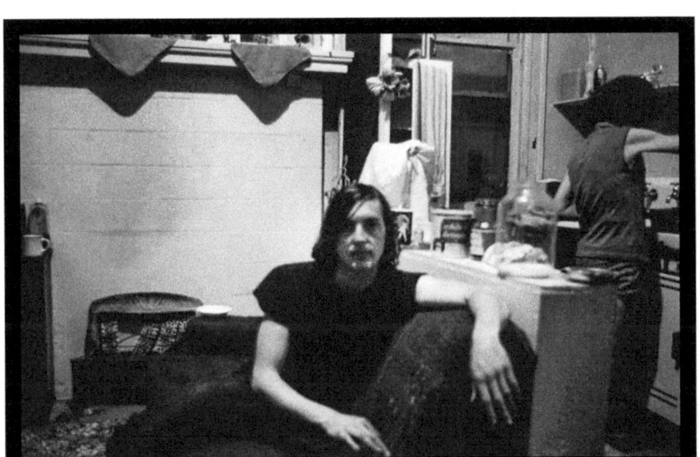

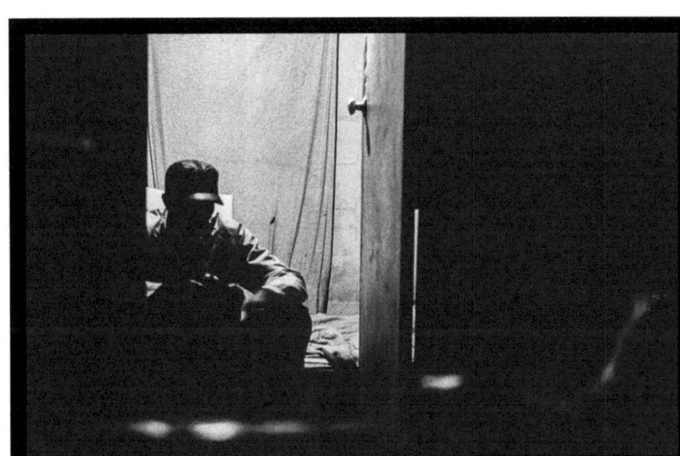

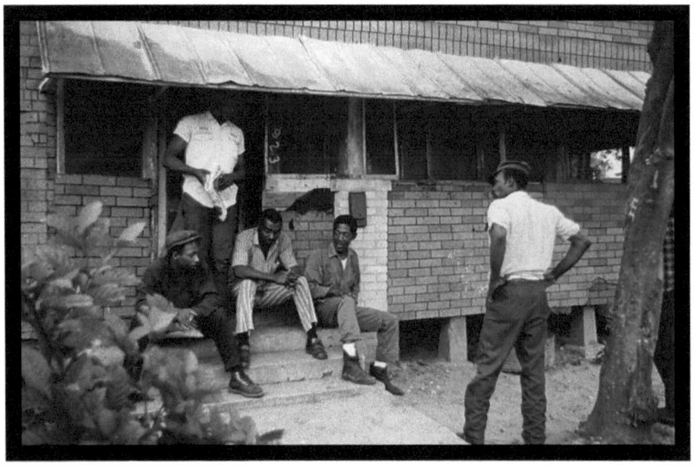

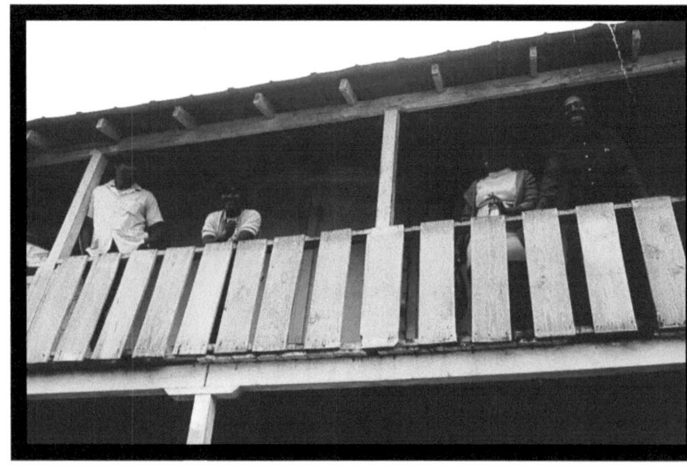

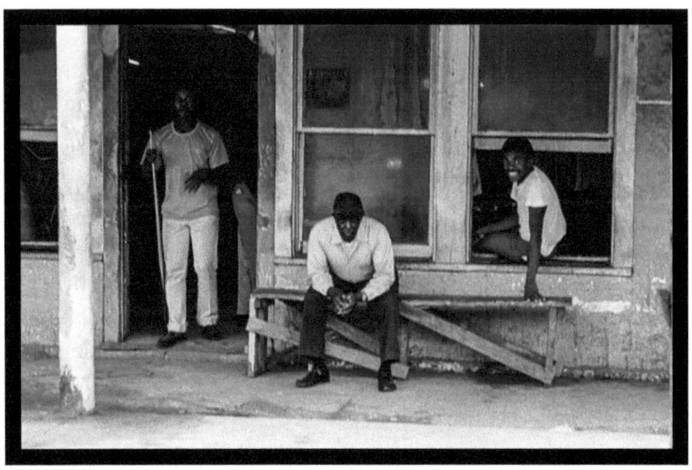

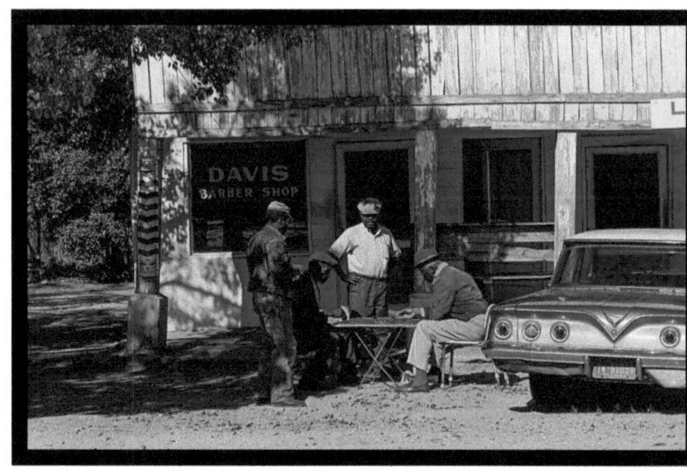

DAVIS
BARBER SHOP

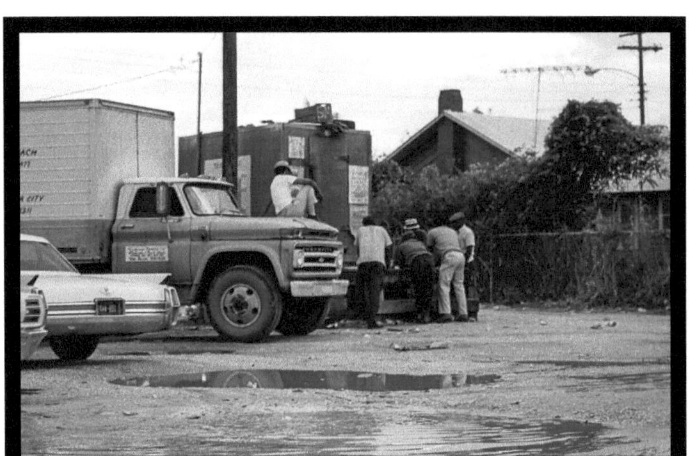

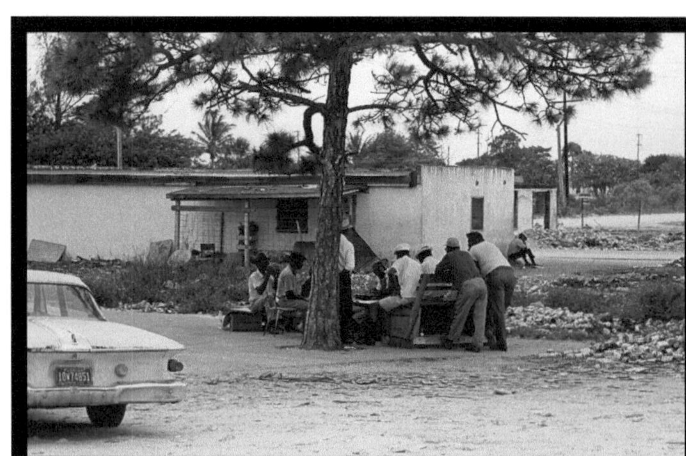

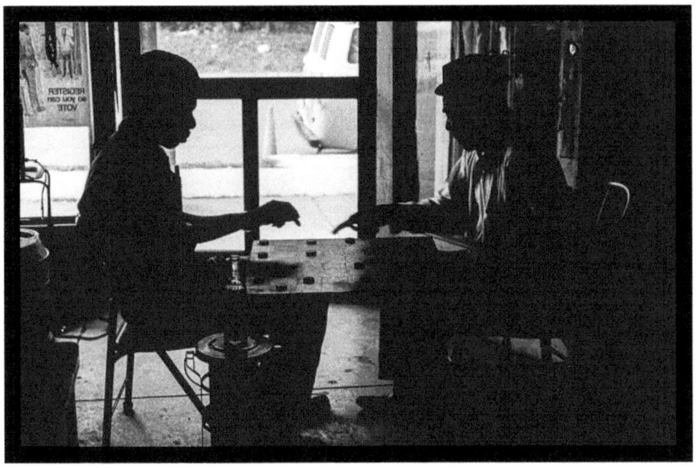

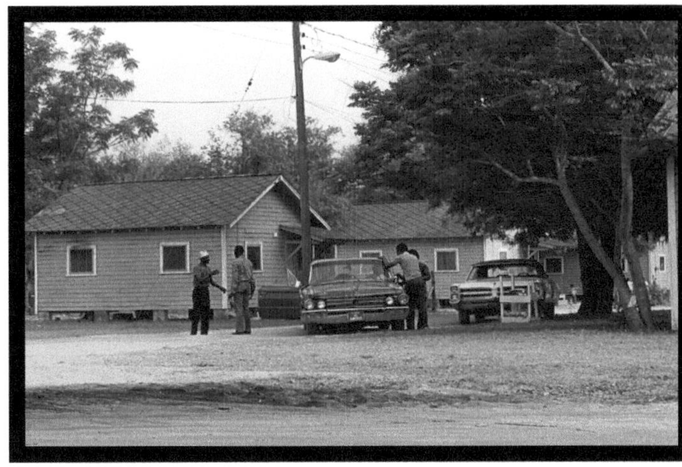

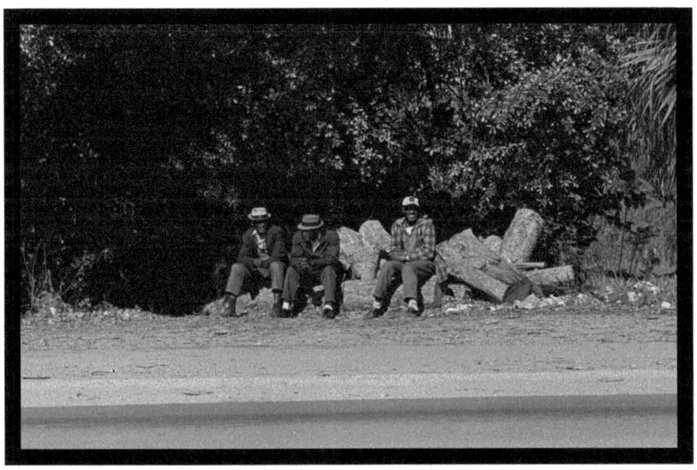

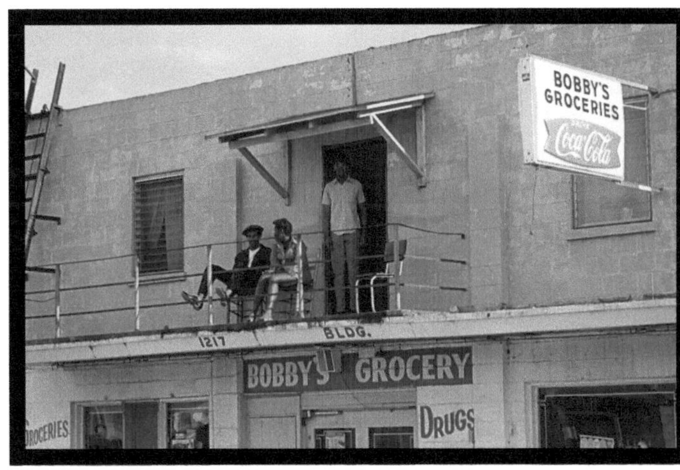

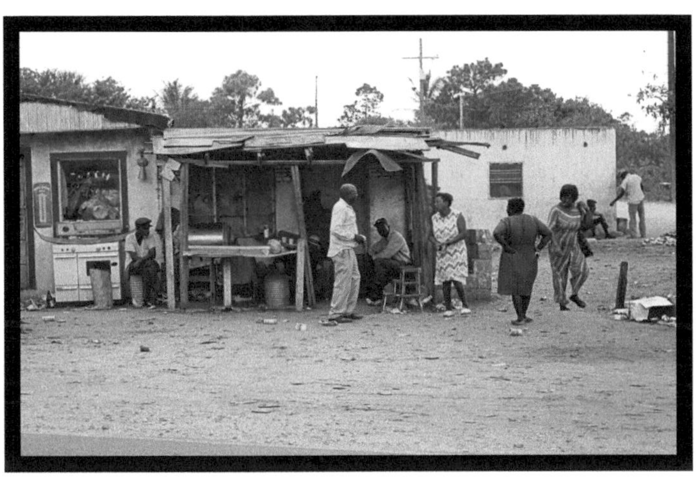

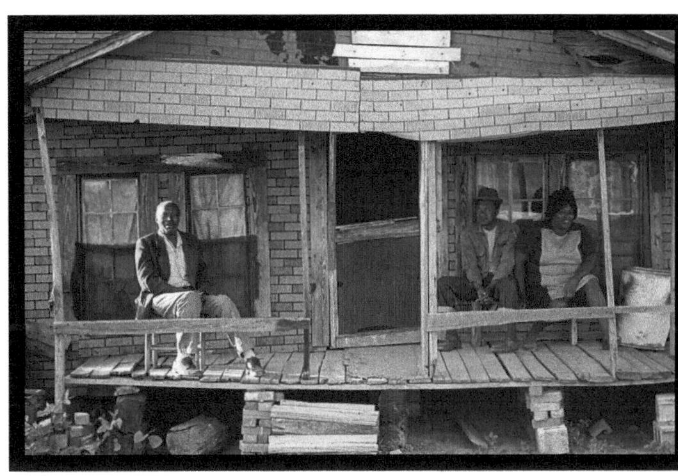

Tom Policano

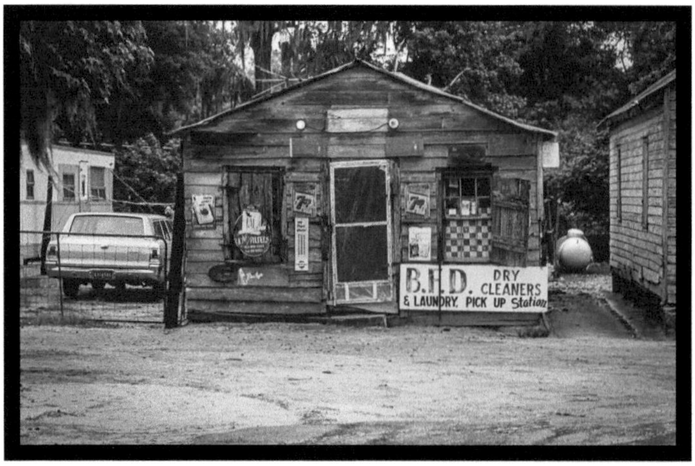

B.F.D. DRY CLEANERS
& LAUNDRY. PICK UP Station

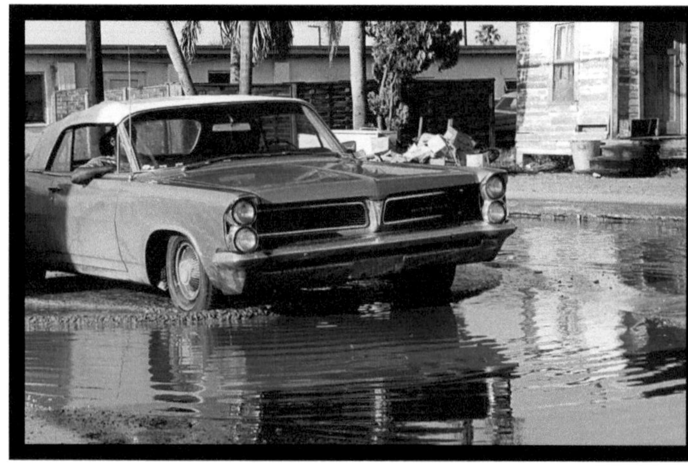

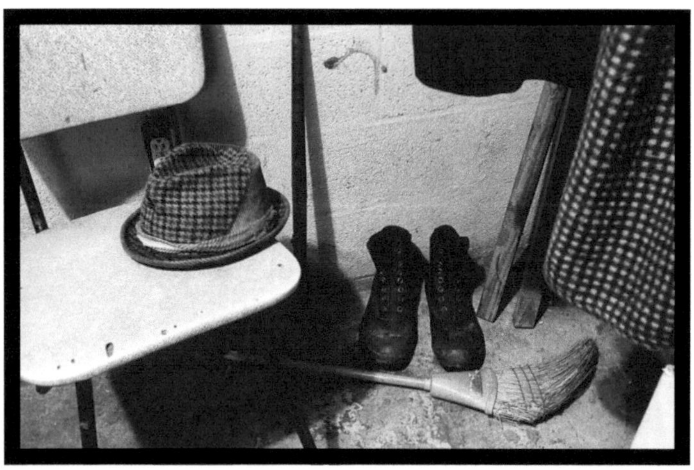

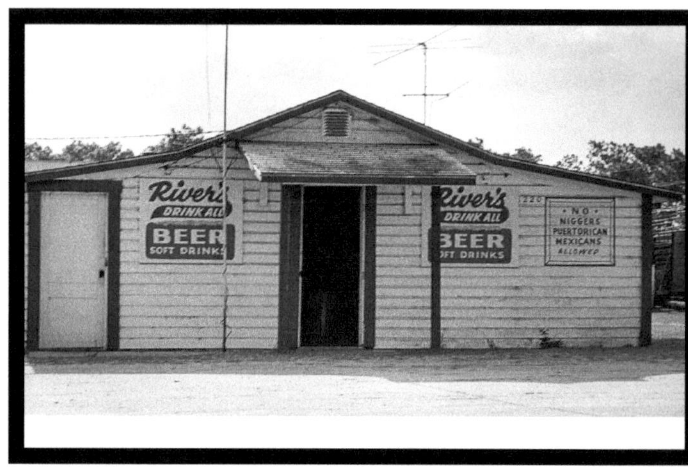

River's DRINK ALL
BEER SOFT DRINKS
River's DRINK ALL
BEER SOFT DRINKS
220
NO NIGGERS PUERTORICAN MEXICANS ALLOWED

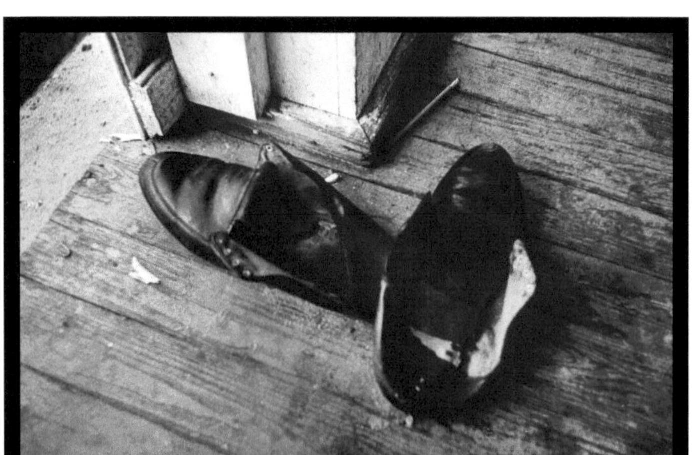

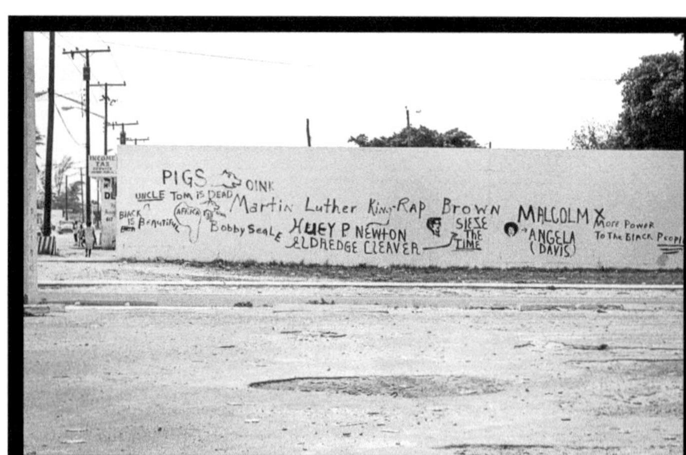

PIGS OINK
UNCLE TOM IS DEAD
BLACK IS Beautiful
AFRIKA
Martin Luther King-Rap Brown
Bobby SeaLE
HUEY P. NEWTON
ELDREDGE CLEAVER
SIESE THE TIME
MALCOLM X
ANGELA (DAVIS)
More Power To The Black People

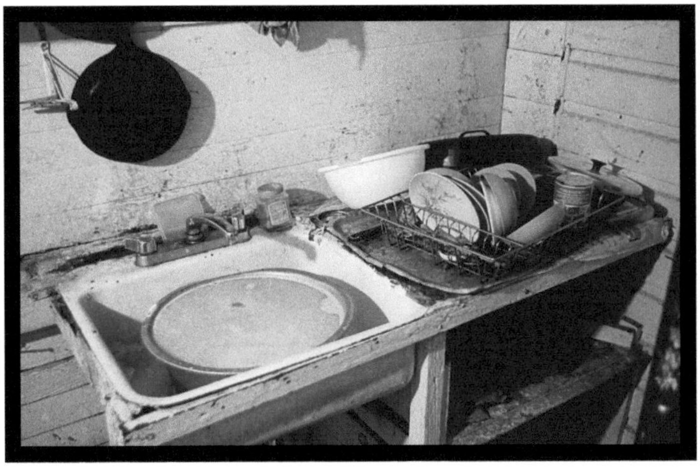

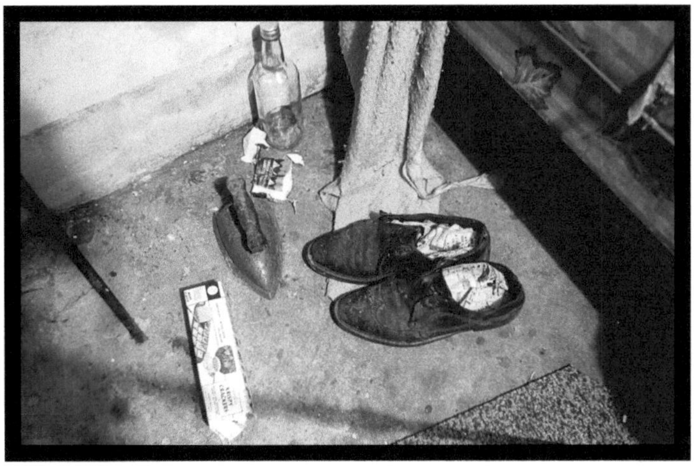

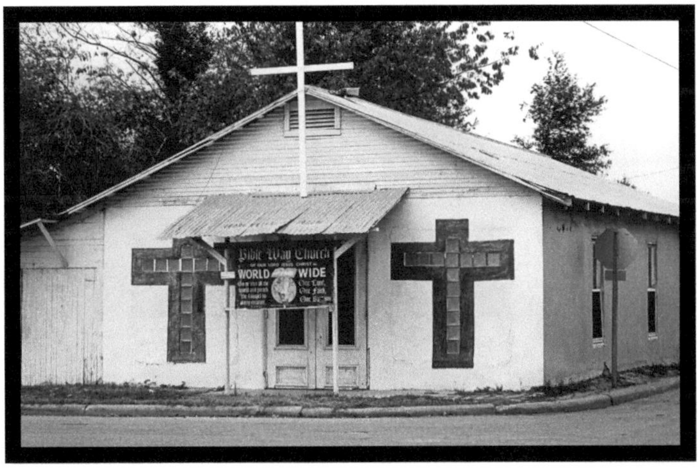
Bible Way Church
WORLD
WIDE

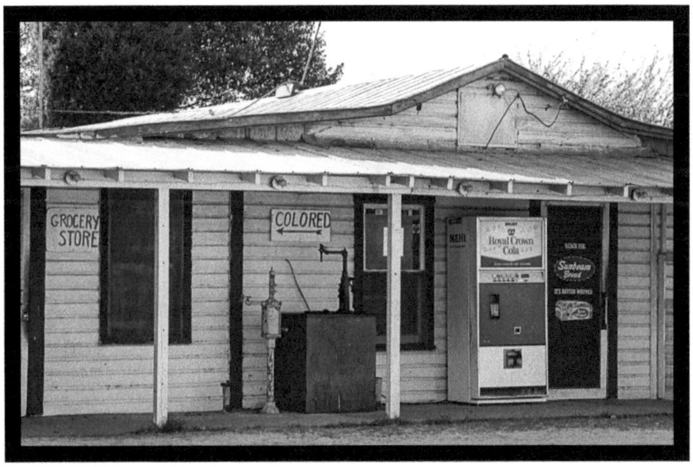
GROCERY
STORE
COLORED
NEHI
Royal Crown
Cola
Sunbeam
Bread

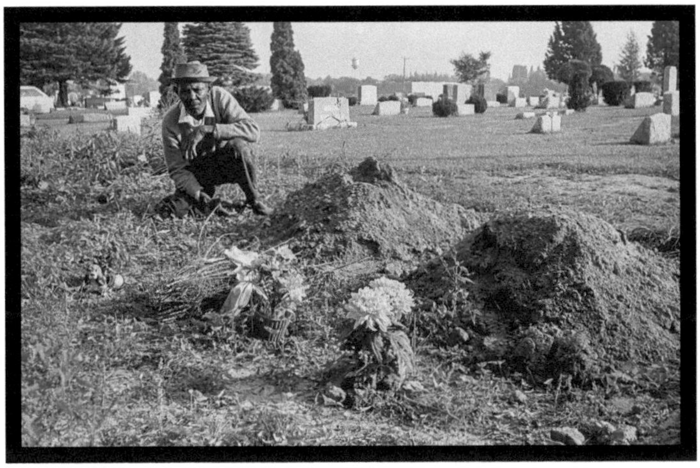

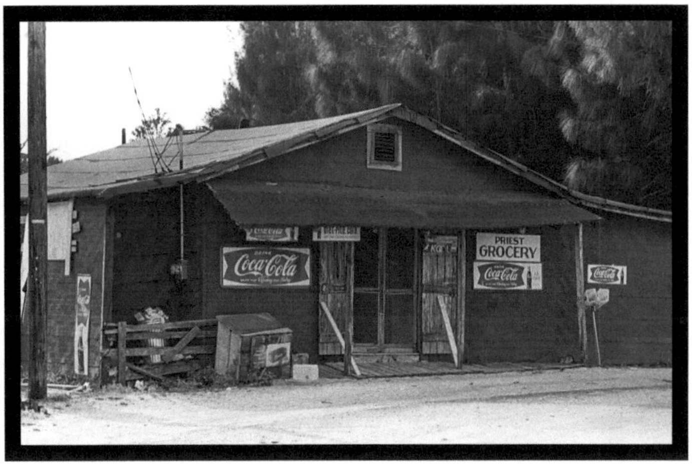
Coca-Cola
PRIEST
GROCERY
Coca-Cola

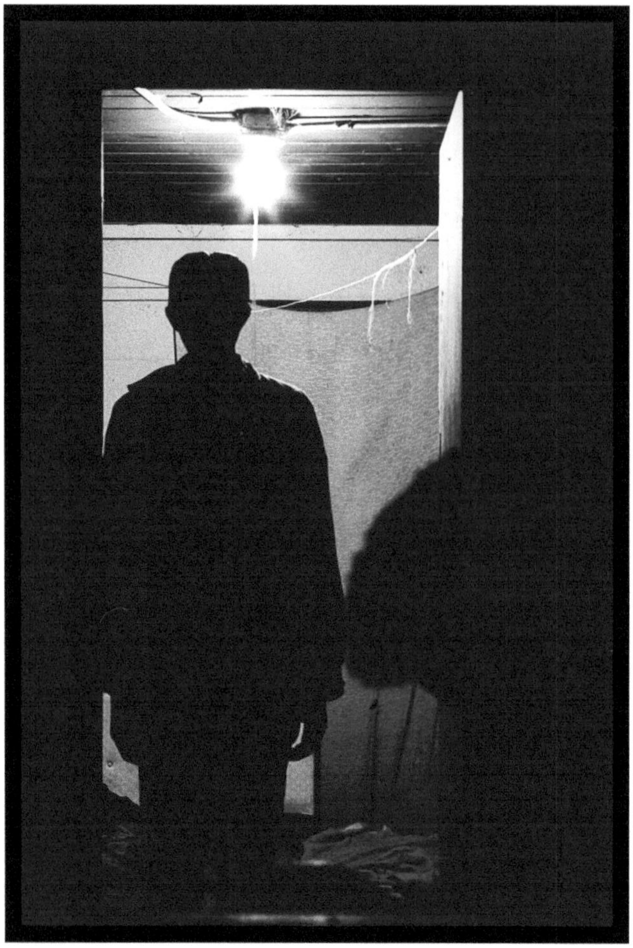

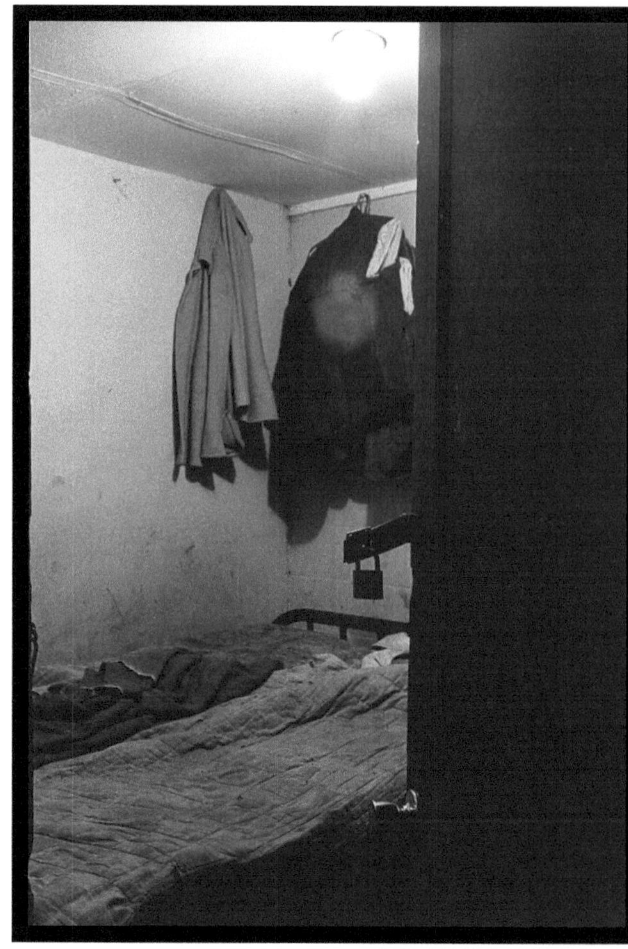

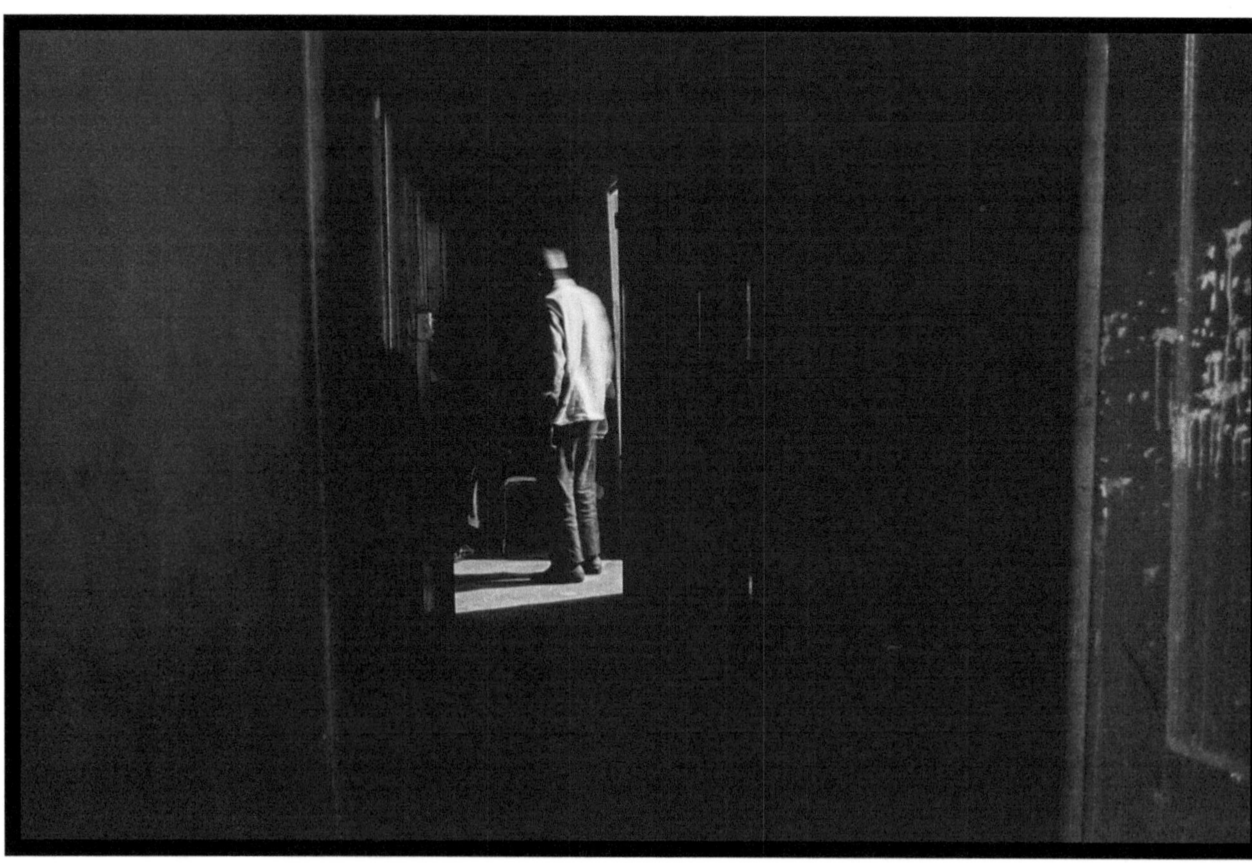

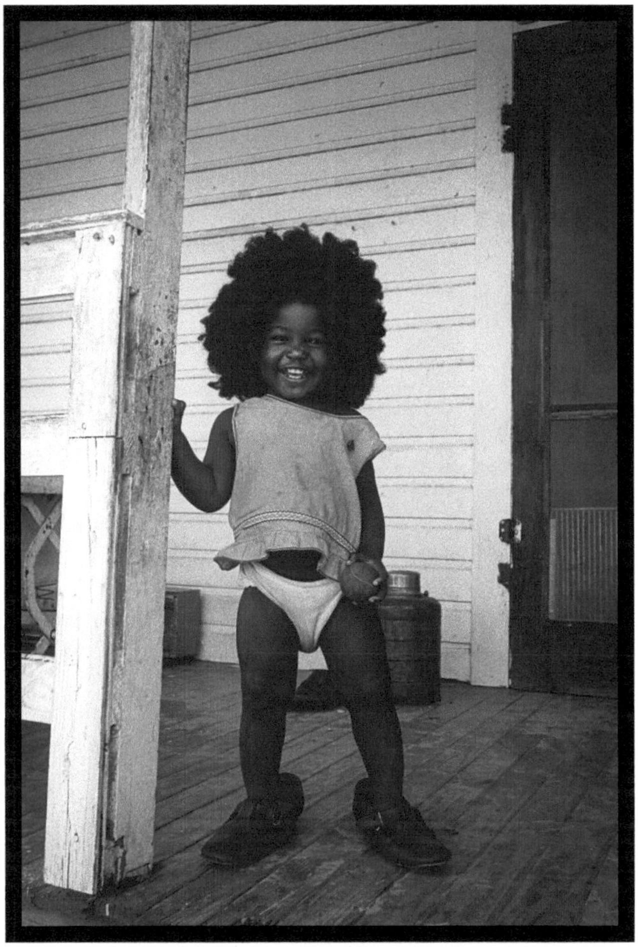

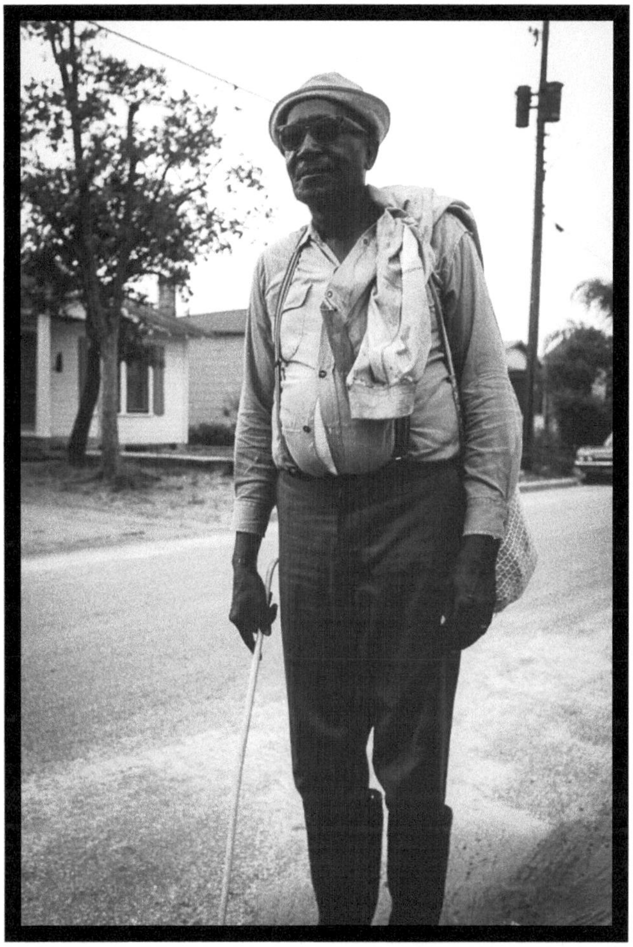

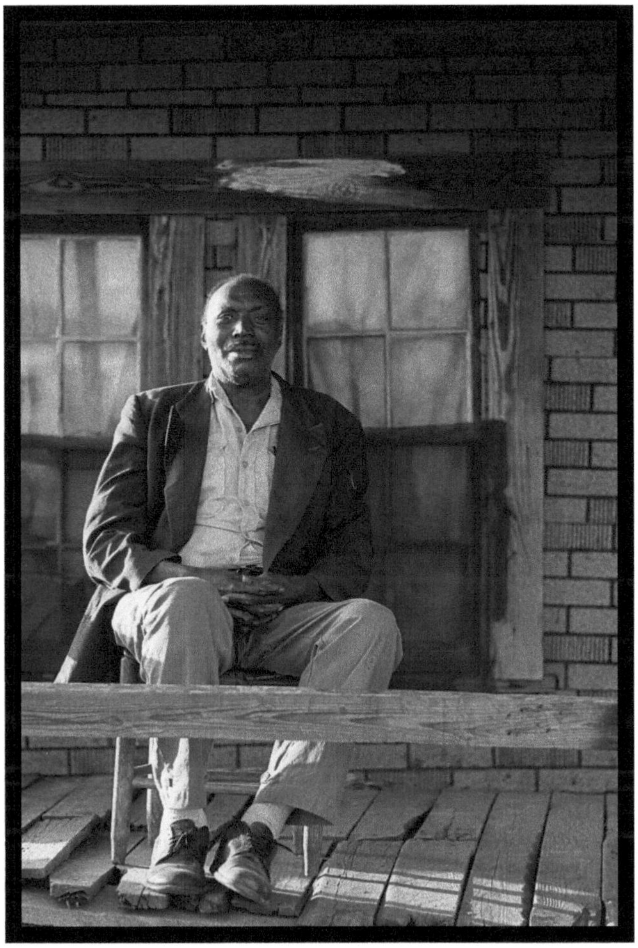

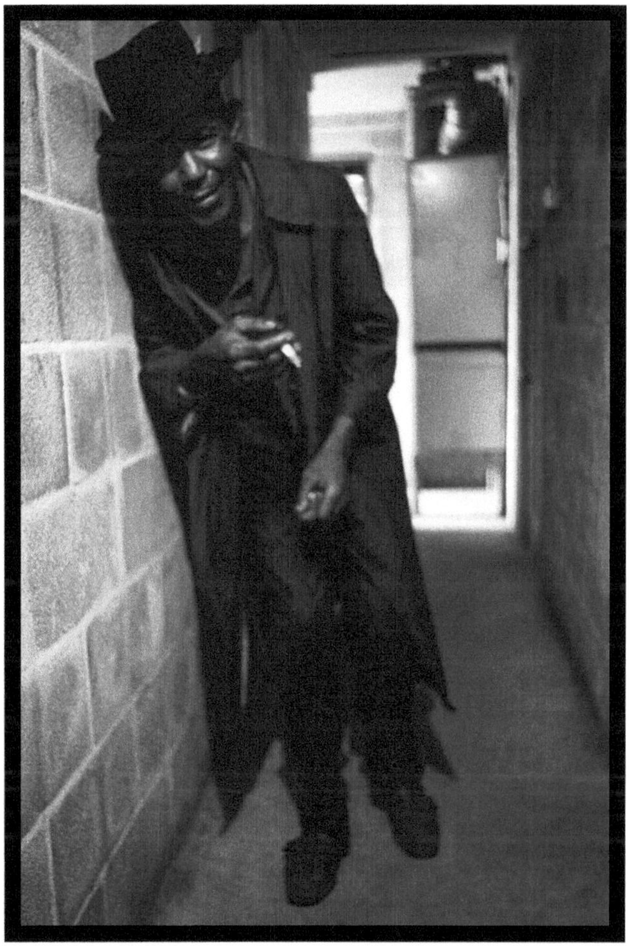

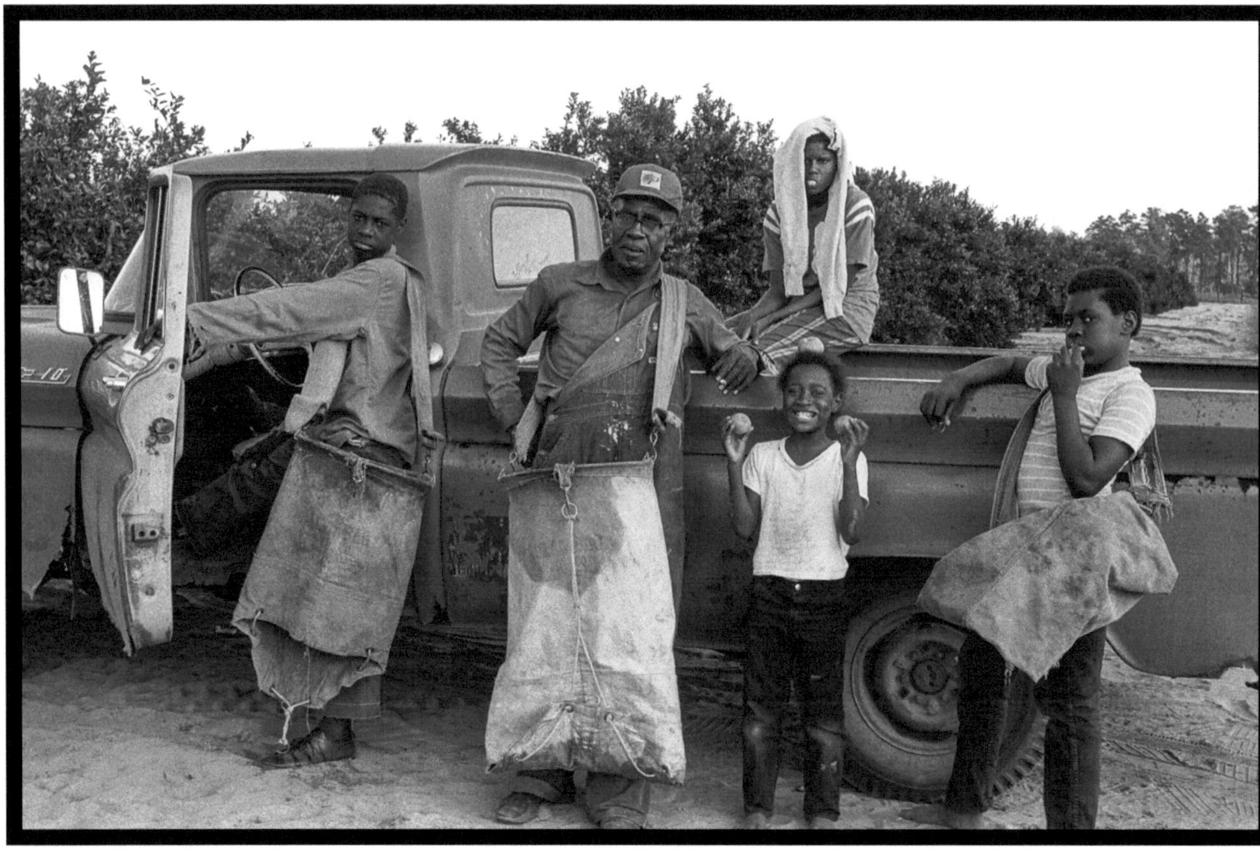

Reverend Peterson

Reverend Peterson was glad to see me. "Why I'll be, Tom, we thought you went back to New York." We sat on the porch and talked quite a while. Reverend was feeling pretty low with only $305 a month coming in. That includes his war veterans' disability check, plus the income his wife and four kids bring home. There's just not enough to make ends meet, having nine kids and a mother besides himself and his wife to support. It costs $110 for $205 of food stamps, $51 for home insurance, $40 on the car, and $50 on his son's car, 16 gasoline, church donations, medicine… It's a struggle just to keep a roof over everyone's head and some grits in their stomachs.

Reverend's been trying to collect his social security. Although only fifty-three, he's overworked his body and is suffering from arthritis in every limb and joint, emphysema, and high blood pressure. He's considered totally disabled by the Veterans Administration. He's very old for a migrant worker, the average age being forty-nine. He needs what he's saved and worked for; he needs his social security now, not when he's sixty-two or sixty-five, if he's fortunate enough to get that old.

Reverend sees it this way: "Social security is for this particular purpose -- if you don't pay it, you don't look for it back; if you pay it, you look for it back when you're entitled to it. An they refuse to give it to you, it seems like they are takin you or robbin you of your rights. See, this is already yours. You made this an had laid this money up in a bank somewhere. Well, who could deny you from drawin it when you bout ready for it? Nobody tells the government when pay day comes, Must I take five cents, ten cents out. They get it whether you want it or not. An when you get disabled, you look for it back. Why not give it back the same way you takin it out? That's what I'm sayin. You don't have to go through a whole lot of red tape to get it. Here I been sittin up over twelve months now waitin on this, an they still deny me. If I wasn't disabled, if I was able to work, don't you think the Veterans' Administration woulda known it since they had more dealin's with me than the social security people? They didn't deny me, an I only served them about three an a half or four years. But I been servin America for all my life."

While Reverand and I sat on the porch talking, the rest of the family was inside watching television. It was a six inch portable tube, sitting on top of the broken console. The tube stayed on most of the day, no one watching it attentively, but occasionally glancing at it between conversations to see what was happening, unless there was a ball game on. Ball games were very popular with Reverend and the kids.

There were too many people in the five room house, crowded with furniture, plants both real and artificial, pictures of religious subjects and a few family portraits, a giant wall clock with matching wall candles, a few school books, old papers, some records piled in the corners, clothes, diapers, knick knacks, dishes. Neala was stretched out waiting her turn because it was pretty much impossible for everyone to eat together.

Here's a list of those living at the Peterson's:

Andrew 10
Maggie 7
Wilbert 22 (just enlisted in the Army)
Artennie 13
Orvern 15
Beverly 19
Robert Lee 17
Curtis 23
Daisy May 27 and her kids
Bubble 7
Sylvia 10
Ivlinnie 1
Junior 5 months
Elmer Lee 3
Neala 12
Minnie and Reverend Robert Peterson

Reverend's mother, Maimy Collier, lived in a trailer in the back yard, without plumbing or electricity except for an extension cord run from the house. When I asked her how I should address her, she said, "I'm Mrs. Collier, C-O-L-L-I-E-R, after my third husband. He was a good man. We didn't have any childs though." Well, Mrs. Collier and I were left pretty much alone with the television that night. Reverend was off to the Florida Farm Workers Organization meeting. Mrs. Peterson was sleeping - she was always tired at night from work at the plant nursery. The kids went to a record hop. Only Minnie and Junior were around, but they fell asleep on the couch. We got to talking about fishing and decided to go the next day.

In the News

From My Diary

I arrived in Lakeland, Florida about eleven o'clock and easily found the Florida Farm Workers' Organization's new headquarters on Memorial Boulevard. I had arranged to meet Newlon Lloyd and Paul Wilson there at noon. Reed Johnson was alone in the office when I got there.

Reed is twenty years old. Through the Brethren Volunteer Service, he was assigned to the Farm Workers Organization for his alternative service and tasked to help organize labor. Reed lives with Paul Wilson, a member of the Polk County Christian Migrant Ministry, who along with Newlon Lloyd, an ex-migrant worker and crew leader, began the Farm Worker's Organization in the Fall of 1970 with Migrant Ministry funds. In view of the church's historical "band-aid policy" toward the disadvantaged, this new attitude of helping the poor help themselves is revolutionary. Paul Wilson explains it like this, "Wouldn't the greatest act of charity be to help a man to help himself?"

Other recent events, too, have motivated farm workers to help themselves. Ten months of severe drought with freezing cold spells, the termination of Project Mainstream, weeks left of embargo before crews would pick Valencia fruit, the legend of Chavez, the national sympathy kindled by mass media headlining "The Plight of the Migrant Worker," and the emergence of strong leadership are all bringing workers together. Local communities and state and federal governments find themselves ill-prepared to resist the now obvious needs of the once helpless, always exploited farm laborer. Throughout Florida, new home of the Black from Dixie and the Chicano from Texas and the Midwest, wherever workers come together, the establishment is yielding and the pride of a once meek people is swelling. Social welfare, family assistance, workman's compensation, unemployment insurance for farm workers is around the corner.

We sat there: Newlon, Paul, Elmer Dixon, an organizer in from Bartow, Reed and myself talking, planning. Newlon commented, "That PA system we got is the best organizing tool we have. We used it last week in Haines City and got three hundred fifty people to our first meeting there, eighty-four of em joining the organization that same night. We're planning a tour after the regional meeting on May twenty-second. Going to have meetings in Orlando, Fort Pierce, Vero Beach, Bartow, all over the Citrus Belt. We'll go in with the PA and have the meetings that same night. Next year we'll be ready to strike if we have to."

Everywhere organizers envision the new tomorrow they can have if they get together. Join us. Organize. ORGANIZE!

I wonder will workers remain interested? Can the people help themselves? Do they want to help themselves? Or will it cost the system a fortune for people-centered community action programs headed by professional poor, concerned primarily with self-improvement? Can poor people, classically generous and compassionate, remain that with the new freedom and responsibility that federal money will give them? Will the leadership, realizing their power, seek only more power and forget to educate the people as they go? Will their actions polarize the groups involved and the nation at large, and would such polarization end in violence?

Tom Policano

All these possibilities flash through my mind. I don't know. These questions and many more come to me out of the experience of being with all kinds of people on all levels of involvement and concern with migrants. I'm doing a social landscape of farm workers on the east coast, I hear myself telling new people every day. I've been working on it a year now. Got my first car three months ago and left New York the next day. In these months I've travelled 11,000 miles. Lived, slept, worked, got drunk with and fought along side of farm laborers. Picked grapefruit in Winter Garden, oranges near Cocoa Beach, tomatoes in Immockalee. Been to organizational meetings in Bartow, Lakeland, Haines City, Immockalee, Palmetto. Offered my services to the leaders of OMICA (Organized Migrants in Community Action) in Imockalee and Wachula, and the Florida Federation Farm Worker's Organization in Bartow and Lakeland.

I was in an Easter parade to bring the Church and the farm labor movement closer together, a march to present demands before a town council to hire two black policemen and begin a policy of more protection and less harassment, on a three hundred and fifty mile trip with fifty-eight other workers in two crew buses to present to the Industrial Relations Board of the State Chamber of Commerce our case for the inclusion of farm workers under workman's compensation and unemployment insurance, and on and on.

I want to tell it like it is, I hear myself saying. Things are changing fast. The migrant worker of the Thirties, with a heritage of slavery, sharecropping, tenant farming and immigration (both legal and illegal) is disappearing. Agribusiness conglomerates now turn to Puerto Rican contract labor to harvest bumper crops, or even establish entire production operations in foreign underdeveloped countries. They are also replacing the domestic labor force with machines that harvest beans, corn, strawberries, potatoes and tomatoes. There is pending Food and Drug Administration approval for the use of an abscission chemical to loosen the fruit so mechanical shakers can be used to harvest oranges, grapefruit, apples, and other tree crops as well.

"Tom, you're coming to the meeting tonight in Haines City, aren't you? Why don't you meet Newlon and I at the Holiday Inn outside Lakeland for dinner," Paul's voice suddenly bringing me back, scattering all those pieces I wanted so very much to put together.

I had never eaten in a Holiday Inn before, although there seemed to be one in every town. You see signs advertising them everywhere. If you're in a Holiday Inn, you know you're in a "safe," reputable place. I would guess the regular clientele are businessmen. The menu was fancy, and so were the prices. Newlon and I got there first and had a long conversation. Paul arrived and we ordered, Newlon the seafood platter, Paul the stuffed shrimp, and I the Greek salad, the cheapest thing I could find at $1.75. The food was badly prepared and Newlon was upset, while I couldn't help wondering why we weren't eating soul food at some small restaurant where you know you'll get good food at a right price.

Paul and Newlon noticed everyone who came in, spotting an important grower, then some pretty women. A while later our black people came in, and Paul quickly brought our attention to them as if the Holiday Inn had just become desegregated and this would be the first test case. After dinner Newlon paid the check, using his Master Charge card.

Haines City hadn't changed much in two months. I remembered asking a gas station attendant just off Route 27 for directions to Avenue I, where Albert Cummings, a worker from New York, lived. He had looked quite amazed: "Avenue I? You know that's in the colored section." When that didn't bother me any, he continued, "Go to the top of the hill, turn left and cross over the tracks. Make a right and follow the tracks on down." Turns out in Haines City black people live on the lettered avenues and white people on the numbered avenues.

It was about seven o'clock. The sun was starting to cast long shadows but was still bright and warm. On the crowded streets and dirt roads the men still gathered on the same porches in front of the same stores. The kids were still playing in the streets. Not many people had left yet to go up north. Most of those I talked with now thought they might, though I wondered how many would decide not to since the work had been so bad last year, and then again how many would go even though it had been bad last year, knowing it would be worse here. The meeting hall was locked, and we had to climb in through a window to open the place up. The meeting was called for seven-thirty, but no matter what time a meeting is called for, people won't show up 'til after dark.

At about nine-thirty, Newlon started the meeting with about a hundred workers present. He spoke about how different towns were being organized and of how the workers in Lakeland were now getting forty and forty-five cents a box for oranges, rather than thirty-five cents, because together they had refused to work for less. He explained that the legislation to extend workman's compensation to farm workers had passed the State Senate, but had been modified by the House to exclude farmers having liability insurance. This meant that an injured farm worker himself would have to hire a lawyer to first prove that he had not been careless and then that the farmer had been at fault.

This kind of accident case takes years in our court system and costs a lot of money to win. Newlon further pointed out that this addition to the original bill was the result of the lobbying efforts of the Florida Citrus Mutual, a grower organization that just happened to sell liability insurance to members and represented over seventy per cent of Florida's agribusiness interests. He promised that the Farm Workers Organization was fighting the modified bill and assured us that the original bill had already been reintroduced in the Senate.

Paul Wilson then took the floor, explaining clearly what dues were for and about plans to provide comprehensive insurance by next year for members. This insurance would cost only nine dollars a month compared to the forty and fifty dollars a month workers are now forced to pay. With mounting vehemence he raised the issue of the administration of commodity foods, one of three programs of project Mainstream. The program provides federal emergency relief to farm workers in five Florida counties declared disaster areas after several abnormal freezes destroyed the citrus crops in these counties. His point was that blacks should be getting the jobs at distribution centers, not hard-nosed, nasty white people. In his anger at the establishment, Paul went so far as to apologize for his own white skin, and at the end of his talk drew loud applause from the workers. There was no doubt, together Newlon and Paul had impressed the workers with something that night — Organize for Power — ORGANIZE!

After the meeting Newlon headed back to Tampa, Paul to Lakeland, and I, carrying the PA system, was off to Winter Garden. John had finally got his thing together. We were marching down to the city council with a petition signed by three hundred and forty workers to demand that two black policemen be hired with equal status as white policemen and that the black community receive greater protection and less harassment from patrolmen.

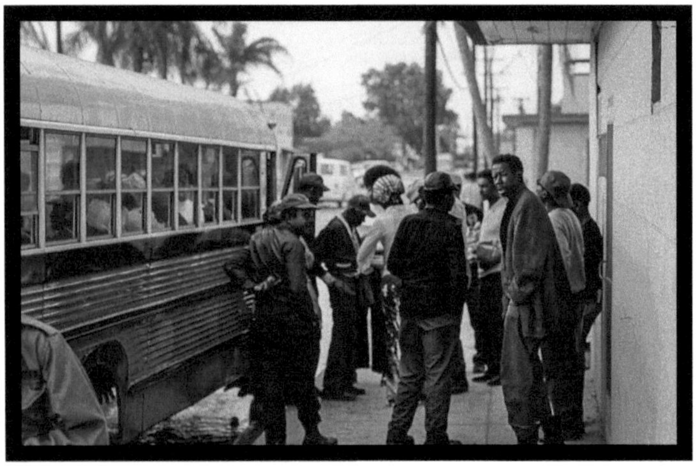

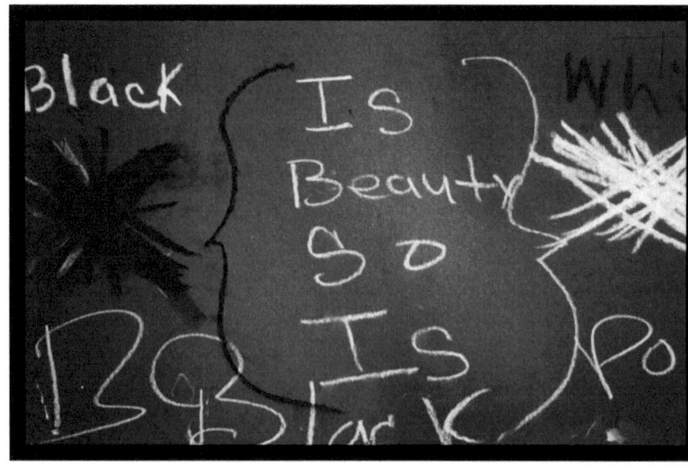
Black
Is
So
Is

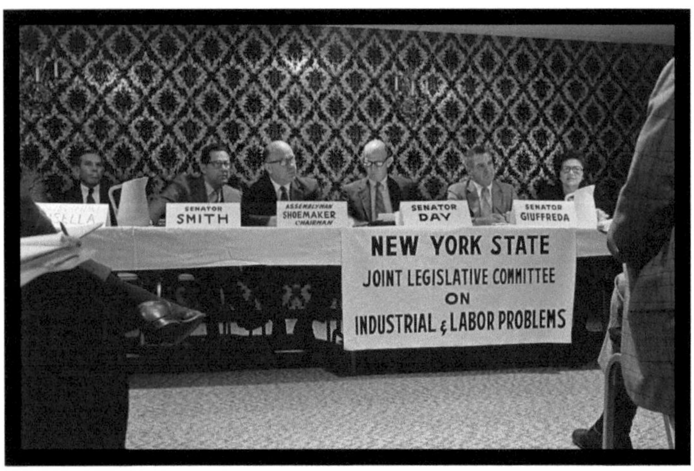
SENATOR SMITH
ASSEMBLYMAN SHOEMAKER CHAIRMAN
SENATOR DAY
SENATOR GIUFFREDA
NEW YORK STATE
JOINT LEGISLATIVE COMMITTEE
ON
INDUSTRIAL & LABOR PROBLEMS

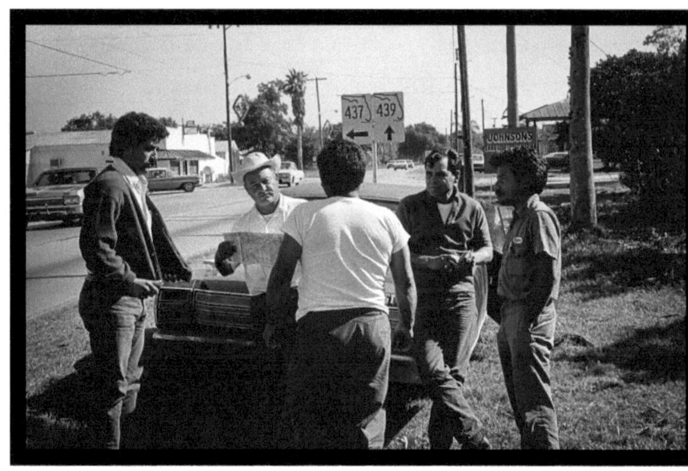
437
439

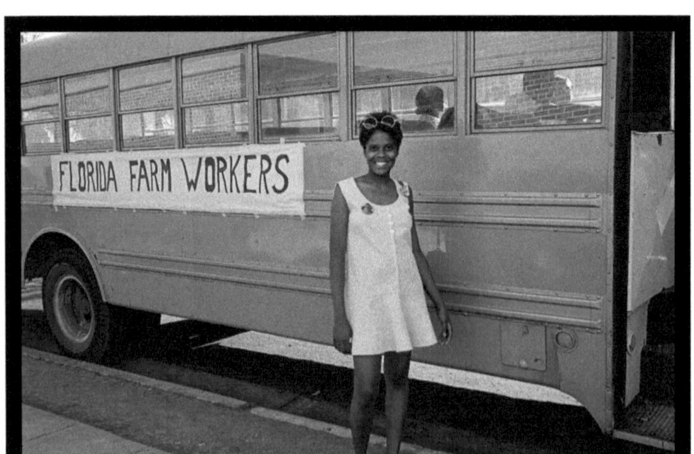
FLORIDA FARM WORKERS

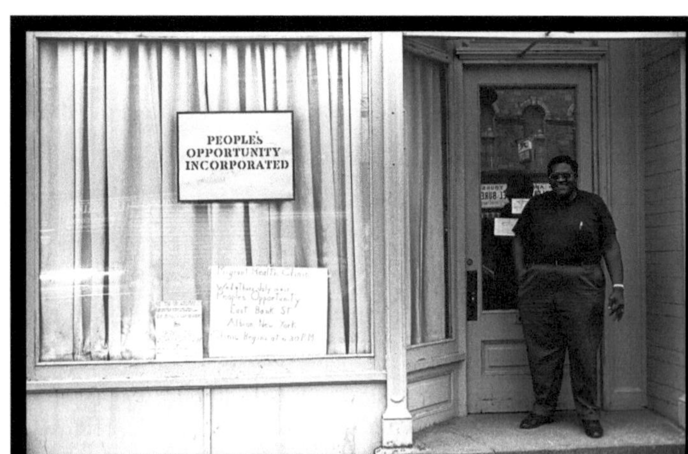
PEOPLE'S
OPPORTUNITY
INCORPORATED

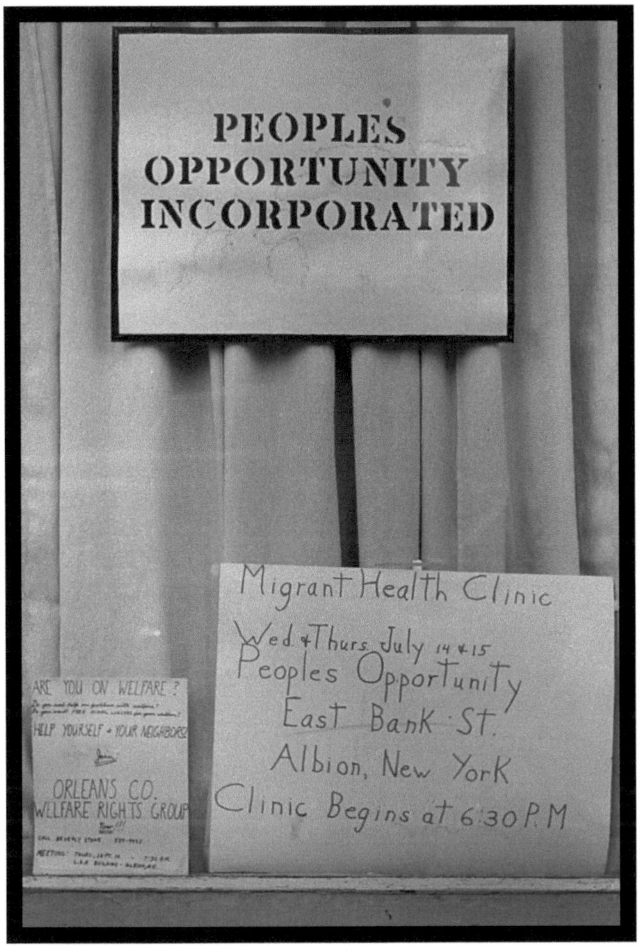

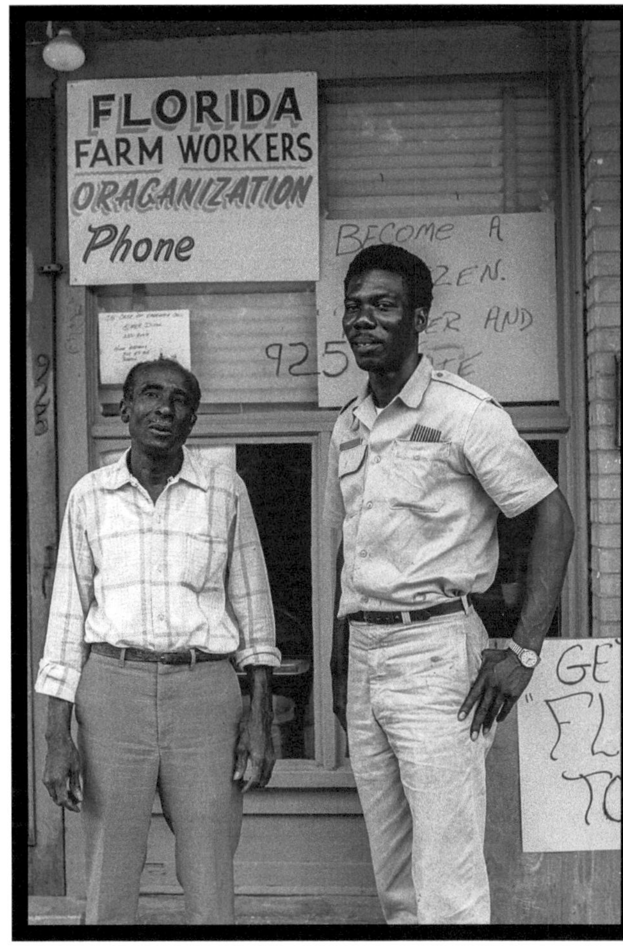

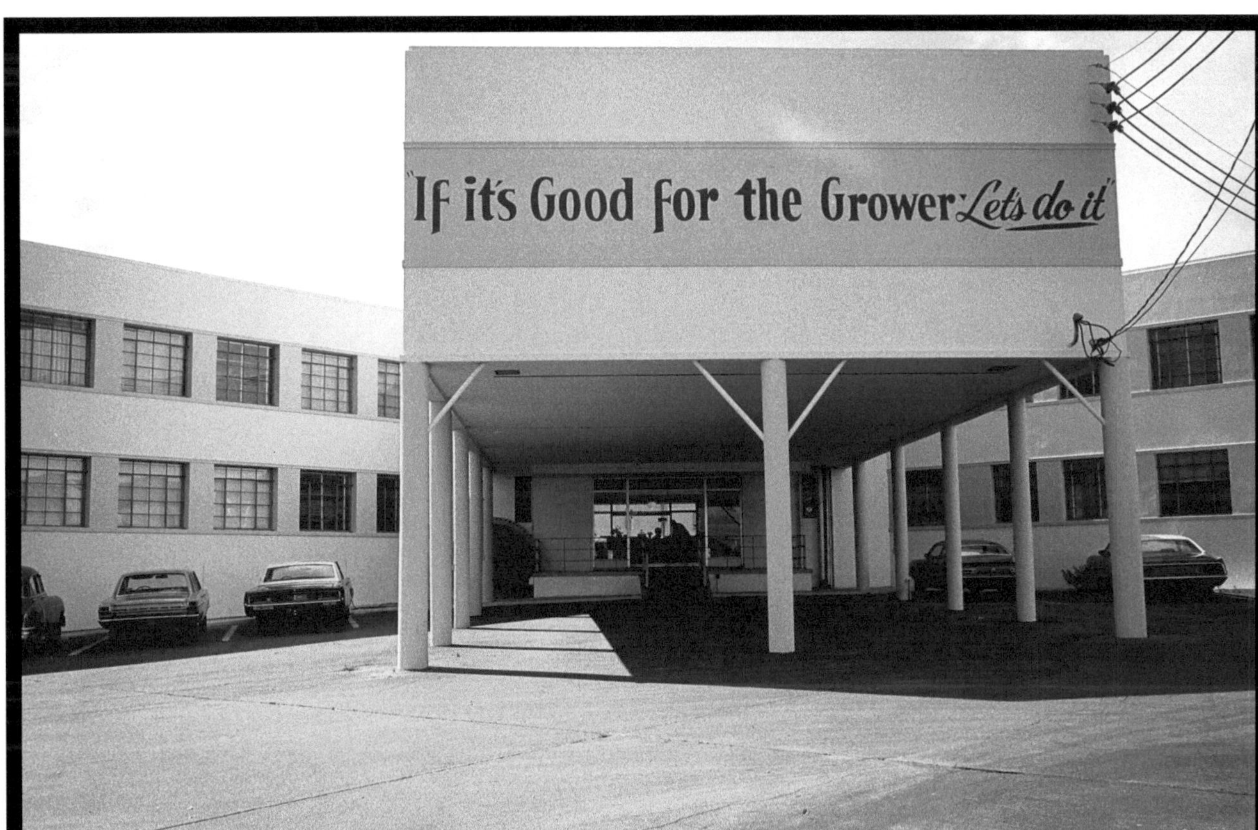

Tom Policano

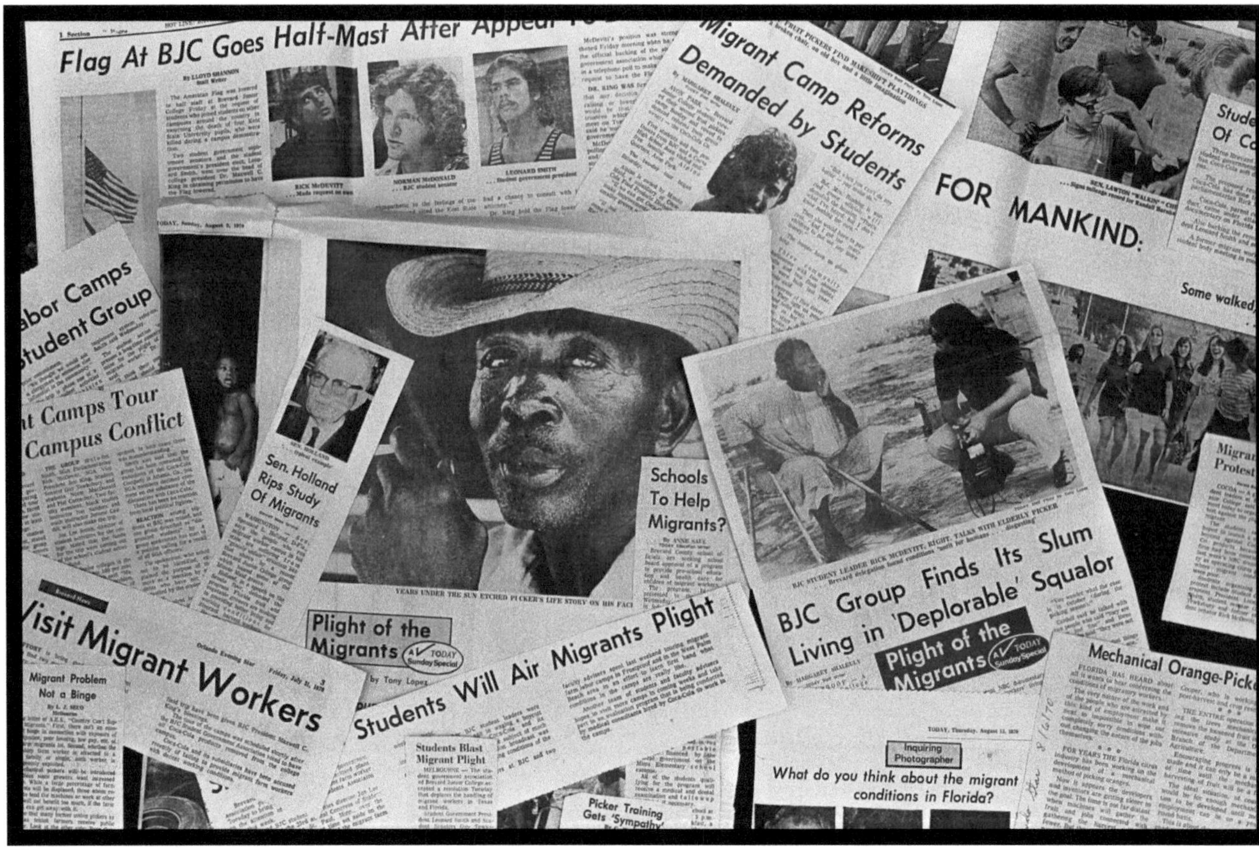
Flag At BJC Goes Half-Mast After Appeal
Migrant Camp Reforms Demanded by Students
FOR MANKIND:
Camps Tour Campus Conflict
Sen. Holland Rips Study Of Migrants
Schools To Help Migrants?
BJC Group Finds Its Slum Living in 'Deplorable' Squalor
Plight of the Migrants
Visit Migrant Workers
Students Will Air Migrants Plight
Mechanical Orange-Pick
Inquiring Photographer
What do you think about the migrant conditions in Florida?
Students Blast Migrant Plight
Picker Training Gets 'Sympathy'

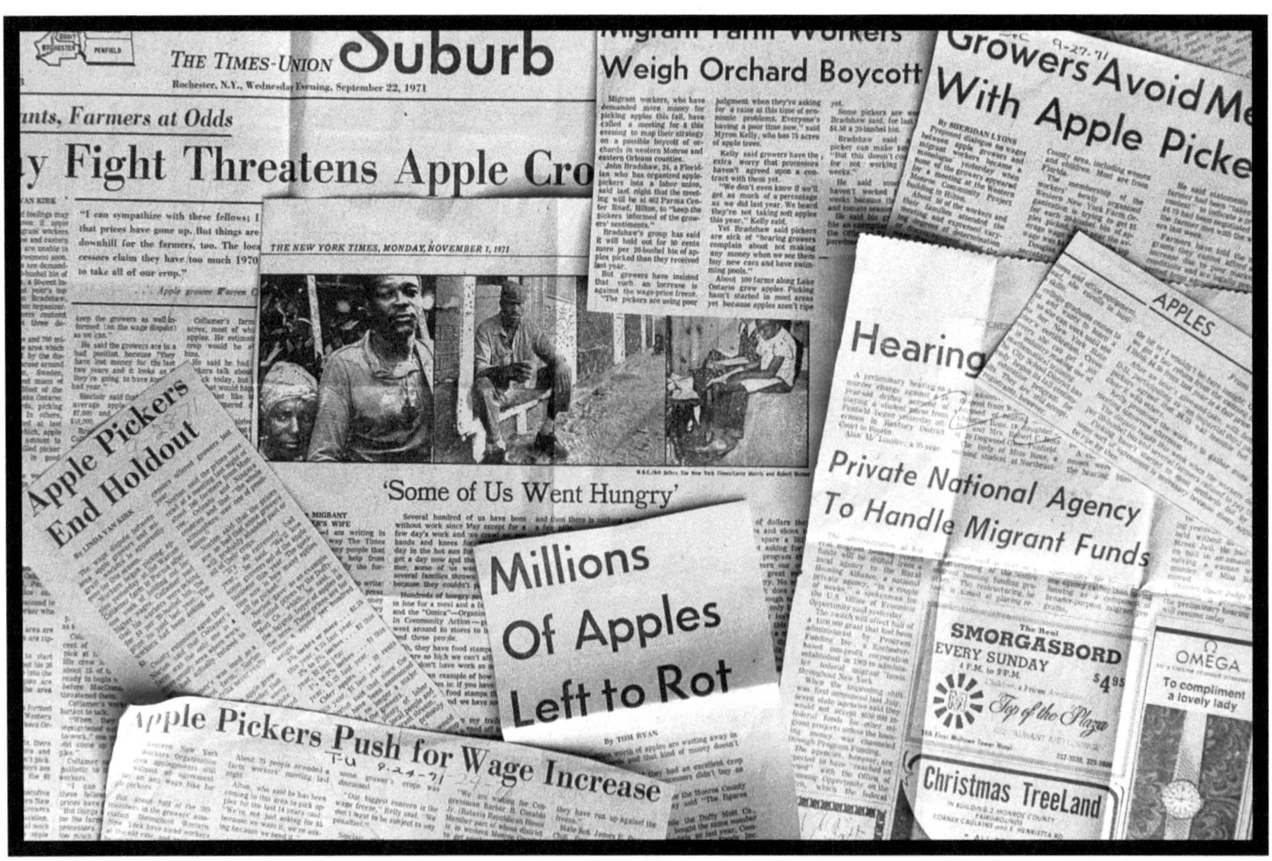
THE TIMES-UNION Suburb
Rochester, N.Y., Wednesday Evening, September 22, 1971
Fight Threatens Apple Cro
Weigh Orchard Boycott
With Apple Picke
THE NEW YORK TIMES, MONDAY, NOVEMBER 1, 1971
'Some of Us Went Hungry'
Apple Pickers End Holdout
Hearing
APPLES
Private National Agency To Handle Migrant Funds
Millions Of Apples Left to Rot
Apple Pickers Push for Wage Increase
SMORGASBORD
EVERY SUNDAY
Top of the Plaza
Christmas TreeLand
OMEGA
To compliment a lovely lady

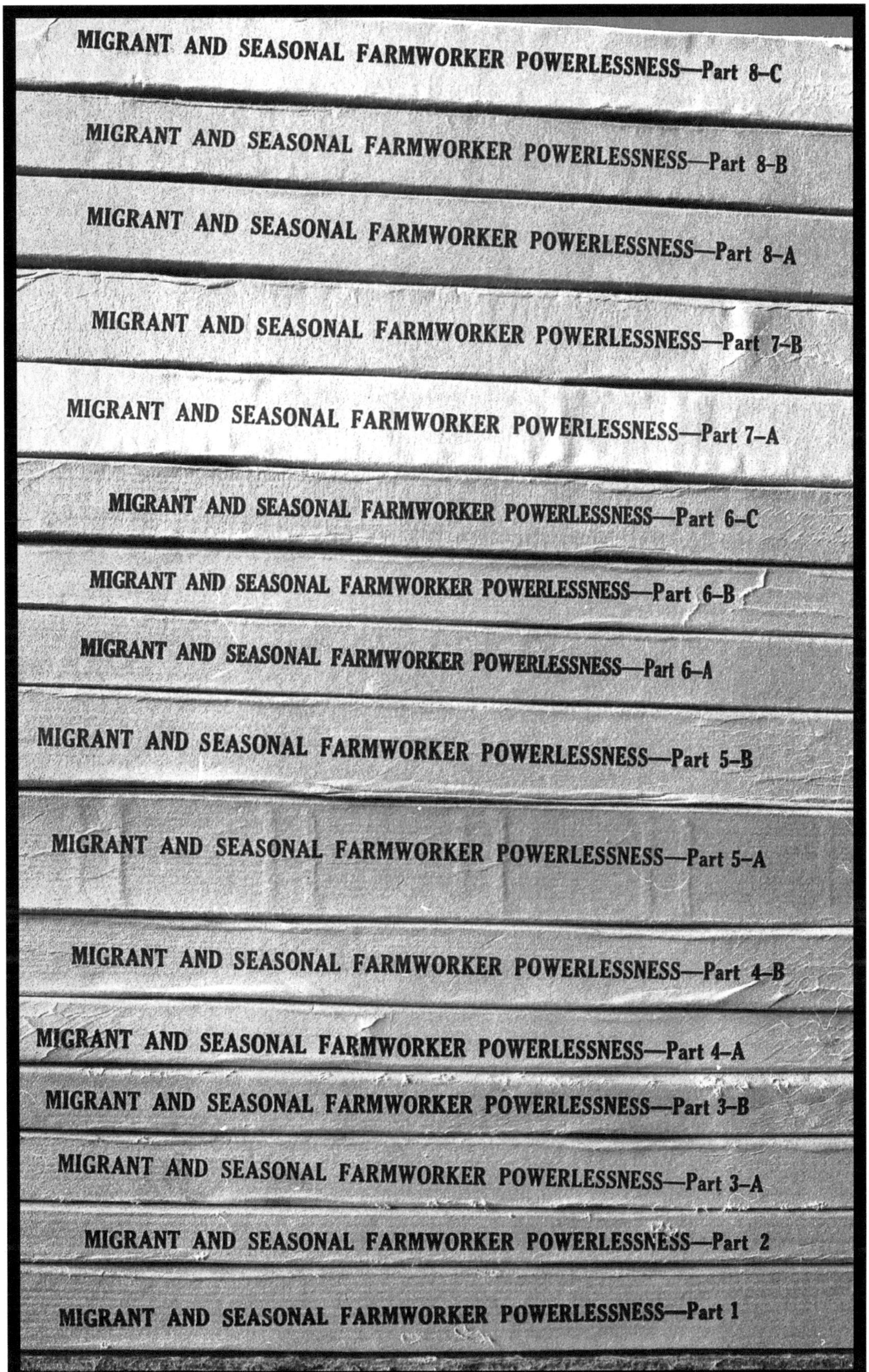

Tom Policano

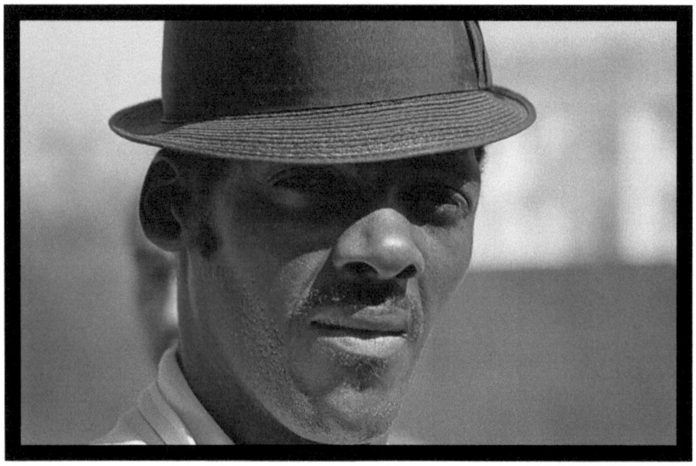
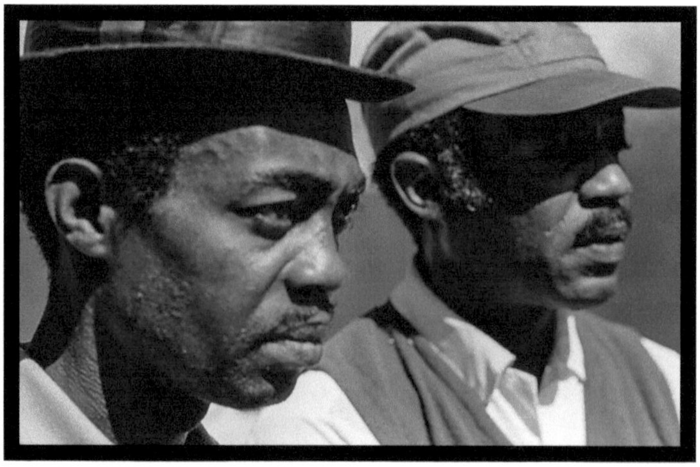
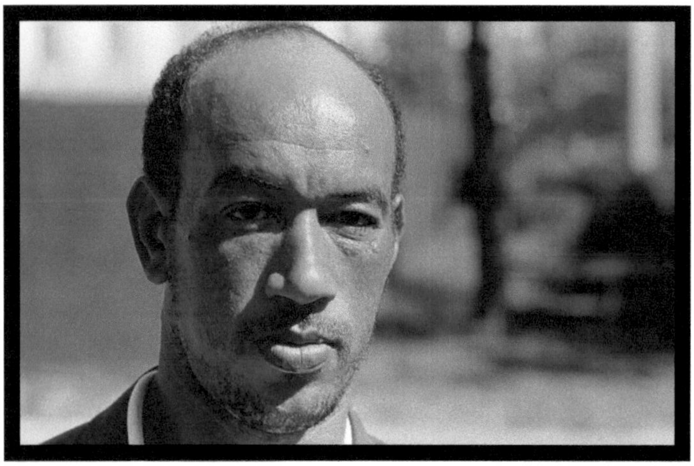
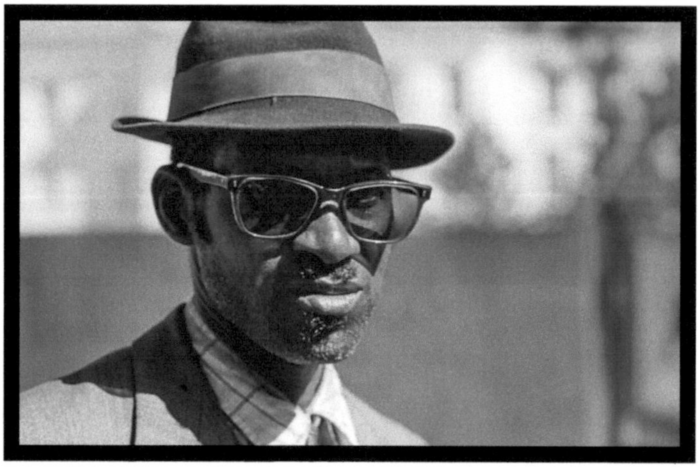
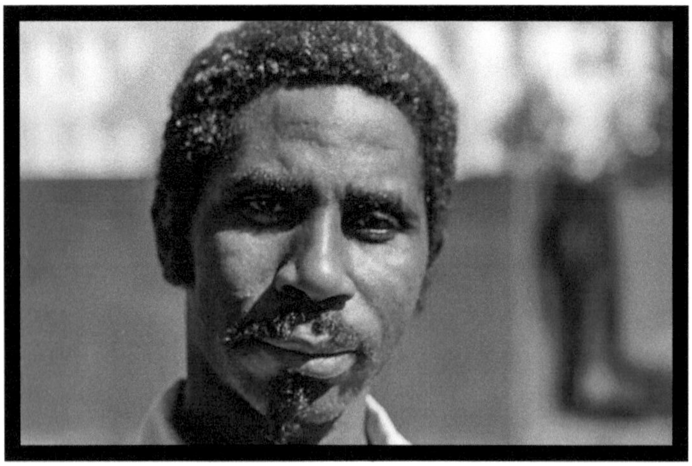
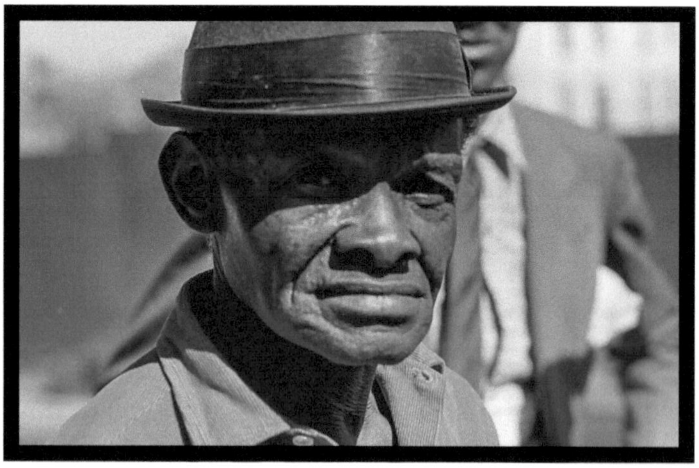

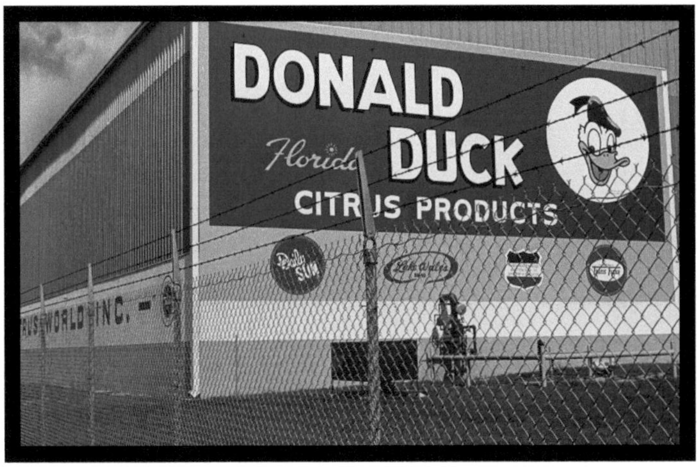
DONALD
Florida DUCK
CITRUS PRODUCTS

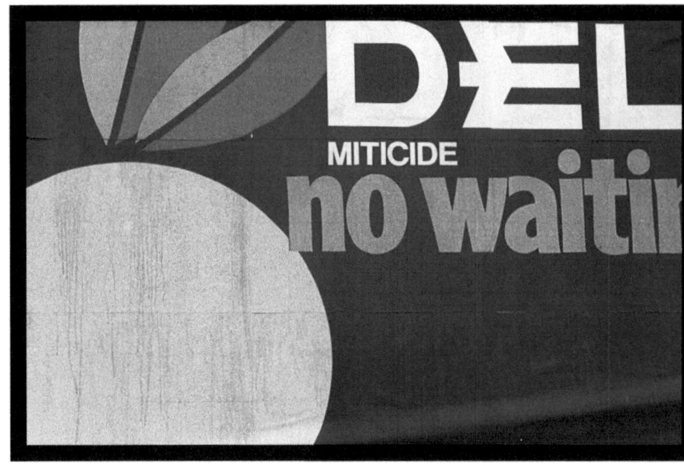
MITICIDE

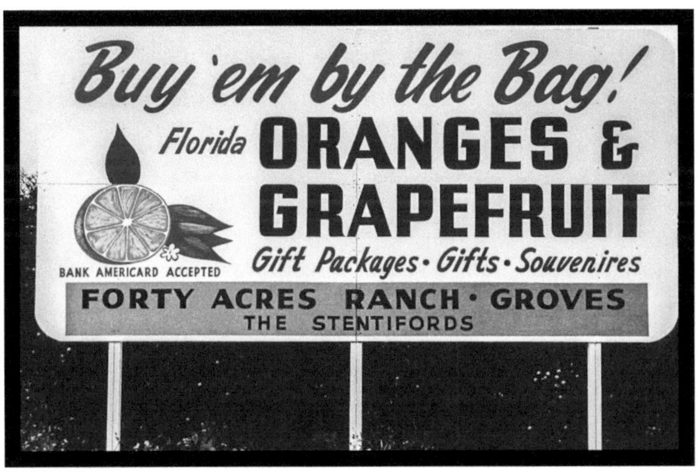
Buy 'em by the Bag!
Florida ORANGES &
GRAPEFRUIT
Gift Packages · Gifts · Souvenires
BANK AMERICARD ACCEPTED
FORTY ACRES RANCH · GROVES
THE STENTIFORDS

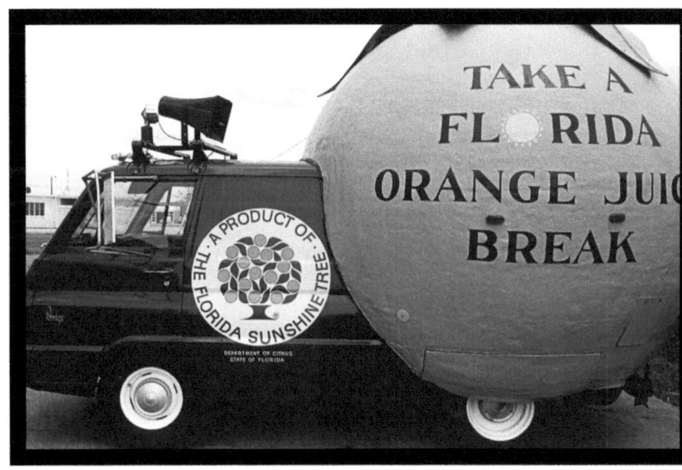
TAKE A
BREAK
A PRODUCT OF THE FLORIDA SUNSHINE TREE

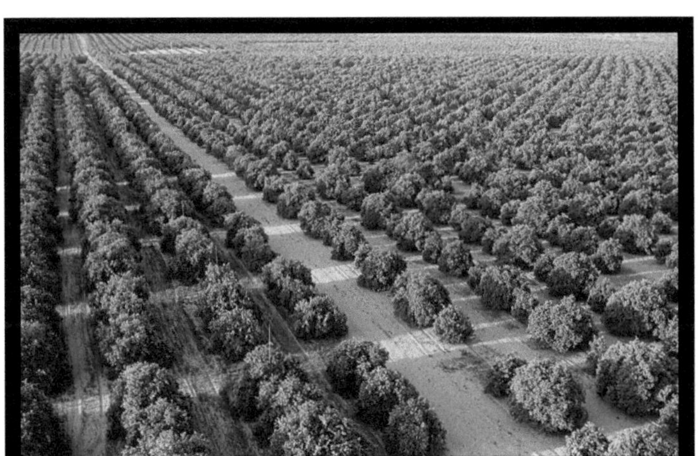

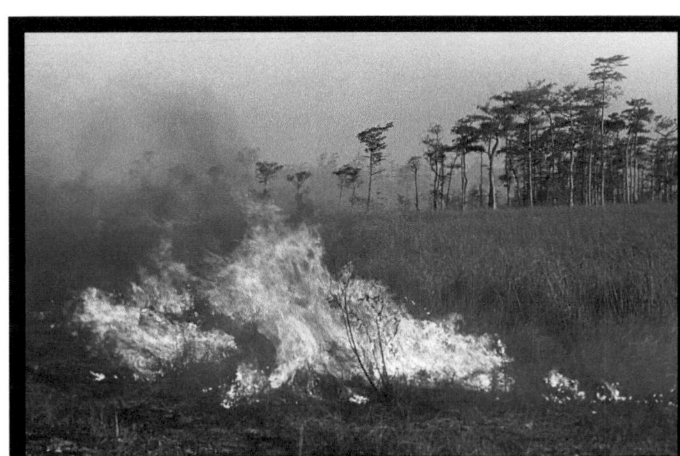

Quotes

I don't think it's easy to justify a system that depends on slave labor, cheap labor that we have to have in order to get those apples picked, or the cherries picked or the potatoes harvested. You know, I think it's hard to justify that system.

They say unskilled migrant labor. This is a fallacy. The migrant laborer is skilled, he's a skilled picker.

The point is that the farmer feels that everybody that enter their camp is after the labor, or he's there to inspect the camp. Why? It should be very obvious because years the cat has passed the buck under the table and he's gotten away with all kinds of things. You know, the inspector has looked around and said fine, beautiful! O.K. Here's your permit.

Let me tell you what I have found. I have found out that the people placed in these key positions are the ones that go along with the government. Right! They aren't for the people. They're political assignments. Right! And, I think, that they should be screened. A person that's going to work with people. Just like these welfare people. They are terrible. They look at you like you dirt.

So you see we're supposed to be doin a lotta stuff. But how do you change so many years? The thing is we have perpetuated a slavery system. Your kids get up old enough to go, you know, maybe to the sixth grade it they're sharp. From six years old to twelve years old. They get out of the sixth grade then that's it. O.K. and they learn just enough to read and write which will be good for the field man out there because he can count his boxes and stuff like that. O.K. and he gets married at about fifteen or sixteen years old because he gets with having to work for his family all the time.

In other words the people in poverty. If you teach them about how to renew with themselves and the world that they live in maybe they'll gain enough knowledge then to help themselves.

Our long range goal although I don't shout it too loud is farm worker unionization. It's the only way that a farm worker has. An'a eventually he'll be automated out of existence and without some sort of effective bargaining agent for the farm workers he's in a big mess.

This is the message I bring for you tonight - organization for power. You must have power to negotiate with your employer effectively. When you begin to speak about organization. When you begin to speak about negotiation. Nobody can negotiate from a weak position. You cannot negotiate effectively from a weak position. You must have something that your employer need and your employer have something that you need. And most of what you have that your employer need Is your labor.

I think it's the beginning of the end.

Epilogue

I used to think I was poor.

Then they told me I wasn't poor, I was needy.

Then they told me it was self-defeating to think of myself as needy, that I was culturally deprived.

Then they told me deprived was a bad image, that I was underprivileged.

Then they told me my underprivileged was overused.

I still don't have a dime, but I have a great vocabulary!

from the *Shelbyville Kentucky Sentinel*

Appendix

Photography notes organized by page from top to bottom and left to right.

Cover	Oct 1971 #09133 Woodworth's Camp, NY	
8	Sep 1971 #08435 Tom Policano, Martin's Camp, NY	
12	Oct 1971 #08905 Granger, crew leader, and Paul reporting a full apple bin	
	Oct 1971 #08901 John Wilcox, apple picking	
13	Oct 1971 #09042 Woodworth's Farm, NY	Oct 1971 #09031 Woodworth's Farm, NY
	Oct 1971 #08936 apples; Woodworth's Farm, NY	Oct 1971 #08910 Woodworth's Farm, NY
	Oct 1971 #08909 Woodworth's Farm, NY	Oct 1971 #08912 apple bin; Woodworth's Farm, NY
14	Sep 1971 #07738 harvesting cucumbers; Woodworth's Farm, NY	Sep 1971 #08322 harvesting cucumbers; Martin's Farm, NY
	Aug 1971 #07434 cucumbers; Woodworth's Farm, NY	Jun 1971 #06938 workers with harvest bushels; Martin's Farm, NY
	Aug 1971 #07429 cucumber collection; Woodworth's Farm, NY	Aug 1971 #07439 Woodworth's Farm, NY
15	Aug 1971 #07436 Levi Cooper taking a break, Woodworth's Farm, NY	Aug 1971 #07437 taking a break; Woodworth's Farm, NY
	Sep 1971 #07728 on the bus; Woodworth's Farm, NY	
16	Aug 1971 #07432 Uncle Tom; Woodworth's Farm, NY	Aug 1971 #07438 Woodworth's Farm, NY
	Jul 1971 #07209 Roland Padilla with co-workers and younger brother	
17	Dec 1971 #09830 asleep on the bus; Palmetto, FL	
	Sep 1971 #08631 Harvesting into the dusk; Martin's Farm, NY	
18	Sep 1971 #08113 Floyd Reese harvesting cabbage, Martin's Farm	Sep 1971 #07724 bushel count by worker; Woodworth's Farm, NY
	Sep 1971 #08105 loading cabbage; Martin's Camp, NY	
19	Sep 1971 #08108 returning to the fields; Martin's Farm, NY	
	Sep 1971 #08110 cabbage; Martin's Farm, NY	
20	Sep 1971 #07735 picking tomatoes; Woodworth's Farm, NY	Sep 1971 #07610 picking tomatoes; Woodworth's Farm, NY
	Sep 1971 #08606 Padilla family picking tomatoes; Martin's Farm, NY	Sep 1971 #08324 Chicano crew with crew leader standing in the middle; Martin's Farm, NY
	Sep 1971 #08233 Juan Padilla picking tomatoes in the rain; Martin's Farm, NY	Sep 1971 #08619 Roland Padilla and brother; Martin's Farm, NY
21	Sep 1971 #07733 picking tomatoes; Woodworth's Farm, NY	
	Aug 1971 #07435 picking tomatoes; Woodworth's Farm, NY	

22	Sep 1971 #07906 Juan Padilla picking tomatoes; Martin's Farm, NY	Sep 1971 #07909 Juan Padilla picking tomatoes; Martin's Farm, NY
	Sep 1971 #08229 picking tomatoes; Martin's Farm, NY	Sep 1971 #08238 Martin's Farm, NY
	Sep 1971 #08236 Martin's Farm, NY.	Sep 1971 #07608 unloading empty tomato bushels; Woodworth's Farm, NY
23	Sep 1971 #08327 picking tomatoes; Martin's Farm, NY	Sep 1971 #07732 picking tomatoes; Woodworth's Farm, NY
	Sep 1971 #07907 picking tomatoes; Martin's Farm, NY	
24	Jun 1971 #06932 harvesting beets; Martin's Farm, NY	Sep 1971 #07723 taking a break; Woodworth's Farm, NY
	Jun 1971 #06931 beet harvest; Martin's Farm, NY	Jun 1971 #06929 beet harvest; Martin's Farm, NY
	Jun 1971 #06937 beet harvest; Martin's Farm, NY	Sep 1971 #07730 lunch break on the bus; Woodworth's Farm, NY
25	Sep 1971 #07726 drinking water; Woodworth's Farm, NY	Jul 1971 #07314 Granger, crew leader assistant; Woodworth's Farm, NY
	May 1971 #06242 James Gassaway in the orchard; Woodworth's Farm, NY	Sep 1971 #07641 Luke; Woodworth's Farm, NY
28	Jul 1971 #07236 early morning; Martin's Camp, NY	Jul 1971 #07333 Martin's Camp, NY
	Jul 1971 #07212 Padilla daughter; Martin's Camp, NY	Sep 1971 #08424a Tom Policano with workers; Martin's Camp, NY
	Sep 1971 #08115 fixing the car; Martin's Camp, NY	Sep 1971 #08039 ixing the car; Martin's Camp, NY
29	Sep 1971 #08424 Roland Padilla; Martin's Camp, NY	Jul 1971 #07332 Martin's Camp, NY
	Jul 1971 #07230 loading up to go to the fields; Martin's Camp, NY	
30	Sep 1971 #08035 Martin's Camp, NY	Sep 1971 #08024 Martin's Camp, NY
	Sep 1971 #08716 Martin's Camp, NY	Sep 1971 #08702 Martin's Camp, NY
	Sep 1971 #08711 Martin's Camp, NY	Sep 1971 #08141 Sauceda family; Martin's Camp, NY
31	Jul 1971 #07206 Mrs. Padilla; Martin's Camp, NY	Jul 1971 #07205 Mr. Padilla; Martin's Camp, NY
	Sep 1971 #08426 Roland Padilla; Martin's Camp, NY	
32	Aug 1970 #03041 Sadona Camp, FL	Jun 1971 #06815 Birches Camp, NY
	Jul 1971 #07238 Martin's Camp, NY	May 1971 #06526 Wachula, FL
	May 1971 #06519 Wachula, FL	Oct 1971 #09424 Fernell; NY
33	May 1971 #06525 Wachula, FL	May 1971 #06521 Wachula, FL
	Farmworker Quote	Oct 1971 #09418 Fernell; NY
	Aug 1970 #02242 Thompson's Camp, NY	Jun 1971 #06818 Birch's Camp, NY
34	Aug 1970 #02442 DeBat's Camp, NY	May 1971 #06640 Lakeland, FL
	Apr 1971 #05722 Rondell Moore	Jun 1971 #06820 Martin's Camp, NY
	Jun 1971 #07016 camp; East Bethany, NY	Jun 1971 #07038 camp; East Bethany, NY

Page		
35	Aug 1970 #02501 DeBat's Camp, NY	
	Aug 1970 #01819 children; Martin's Camp, NY	
36	Jul 1971 #07316 Woodworth's Camp, NY	Oct 1971 #09210 payday; Woodworth's Camp, NY
	Jul 1971 #07112 Woodworth's Camp, NY	Jul 1971 #07111 Woodworth's Camp, NY
	Sep 1971 #07643 Levi Cooper barbecuing; Woodworth's Camp, NY	Oct 1971 #09121 kitchen commons; Woodworth's Camp, NY
37	Aug 1971 #07515 Sunday day off; Woodworth's Camp, NY	
	Jul 1971 #07115 Sunday day off; Woodworth's Camp, NY	
38	Oct 1971 #09331 hanging out after work; Woodworth's Camp, NY	
	Oct 1971 #09133 hanging out after work; Woodworth's Camp, NY	
39	Oct 1971 #09338 hanging out Sunday; Woodworth's Camp, NY	
	Oct 1971 #08908 smoking; Woodworth's Camp, NY	
40	Aug 1970 #01223 Reverend Peterson's grandkids; Monroe Co, NY	Jul 1971 #07103 Woodworth's Camp, NY
	Jul 1971 #07109 Woodworth's Camp, NY	Aug 1971 #07418 Woodworth's Camp, NY
	Jul 1971 #07108 Woodworth's Camp, NY	Sep 1971 #08525 Charles & Cathy; Woodworth's Camp, NY
41	Aug 1971 #07420 Claudia and child; Woodworth's Camp, NY	Jul 1971 #07117 Woodworth's Camp, NY
	Jul 1971 #07110 Woodworth's Camp, NY	
42	Jul 1971 #07107 Woodworth's Camp, NY	Sep 1971 #07843 Roosevelt; Woodworth's Camp, NY
	Sep 1971 #07801 Luke & President Scott; Woodworth's Camp, NY	
43	Apr 1971 #05726 Rondell Moore; NY	
	Apr 1971 #05629 Charles McClure; Woodworth's Camp, NY	Apr 1971 #05630 Rondell Moore; Woodworth's Camp, NY
44	Sep 1971 #08522 Roosevelt; Woodworth's Camp, NY	Sep 1971 #08515 Uncle Tom; Woodworth's Camp, NY
	Aug 1971 #07421 Woodworth's Camp, NY	Aug 1970 #01149 Reverend Peterson; Monroe County, NY
45	Sep 1971 #07604 Mackelby; Woodworth's Camp, NY	Aug 1970 #02402 crew leader; DeBat's Camp, NY
	Aug 1971 #07530 Cathy Jones, ex-farmworker; Woodworth's Camp, NY	Sep 1971 #07802 President Scott; Woodworth's Camp, NY
46	Jul 1970 #00626 Sutter family; Orleans County, NY	
	Aug 1971 #07516 Roosevelt with camp dog	Sep 1971 #07640 Luke with camp dog
47	Jul 1970 #00723 Sutter father and son; Orleans County, NY	
50	Mar 1971 #04434 Ernest Battle; FL	Nov 1971 #09608 farmworker crew; Wachula, FL
	Dec 1971 #10917 Lakeland, FL	Dec 1971 #10638 John Wilcox & Claudia with son Isaac; Lakeland, FL
	Dec 1971 #11834 picking oranges; FL	Dec 1971 #10906 Lakeland, FL

51	Dec 1971 #11515 Ninous Isom, crew leader getting bin count from workers	
	Nov 1971 #09607 farmworker crew; Wachula, FL	
52	Mar 1971 #04103 grapefruit harvest; Winter Garden, FL	Dec 1971 #10430 John Wilcox & Claudia with son Isaac; Lakeland, FL
	Dec 1971 #11521 Ninous Isom, crew leader on payday	Dec 1971 #10728 Lakeland, FL
	Dec 1971 #11339 Moses, Wilcox and other workers in van; Lakeland, FL	Dec 1971 #11835 resting; Lakeland, FL
53	Apr 1971 #05912 posted sign; FL	
	Dec 1971 #10441 crew meeting up with farmer, early morning; Lakeland, FL	
54	Dec 1971 #09836 Uncle Tom and crew onboard bus; Palmetto, FL	Dec 1971 #10942 Claudia barefoot on ladder picking oranges; Lakeland, FL
	Dec 1971 #09905 Dead Eyes rolling a cigarette	Dec 1971 #11212 fixing the crew bus; Lakeland, FL
	Dec 1971 #11321 Moses wastes time because of equipment failure without pay; Lakeland, FL	Dec 1971 #11330 Moses waiting for equipment to be fixed; Lakeland, FL
55	Dec 1971 #10902 Lakeland, FL	Dec 1971 #10904 John Wilcox & Claudia with son Isaac
	Dec 1971 #11831 picking oranges	
56	Dec 1971 #11512 broken goat haults work (without compensation); Lakeland, FL	Dec 1971 #10837 killing time playing cards; Lakeland, FL
	Dec 1971 #10400 killing time playing cards; Lakeland, FL	Dec 1971 #11241 killing time; Lakeland, FL
	Dec 1971 #11203 Wilcox, Claudia & Isaac killing time; Lakeland, FL	Dec 1971 #10840 killing time playing cards; Lakeland, FL
57	May 1971 #06129 lunchtime; Immokalee, FL	
	Nov 1971 #09739 smoking a cigarette; Palmetto, FL	
58	Dec 1971 #11517 Ninous Isom, crew leader getting worker bin counts; Lakeland, FL	Dec 1971 #11909 in the orange grove; Lakeland, FL
	Dec 1971 #10905; Lakeland, FL	Nov 1971 #09602 Tom Policano with farmworker crew; Wachula, FL
	Dec 1971 #10434 fixing a ladder while on the job without pay; Lakeland, FL	Mar 1971 #04438 John picking oranges
59	Dec 1971 #10422 Wilcox & crew; Lakeland, FL	
	May 1971 #06132 farmworker day crew onboard bus; Immokalee, FL	
60	Dec 1971 #10428 Lakeland, FL	Dec 1971 #11833 harvesting goat in orange grove; Lake Wales, FL
	Dec 1971 #10809 John Wilcox; Lake Wales, FL	May 1971 #06231 Lake Wales, FL
61	Oct 1971 #08906 Roosevelt	Dec 1971 #11137 farm worker tally sheet, workers are paid by numbers of containers of fruit picked; Lakeland, FL.
	Dec 1971 #10810 Isaac in the grove as parents are working; Lakeland, FL	

62	Nov 1971 #09700 Roberta Frasier; Palmetto, FL	Apr 1971 #06035 loading vegetables; Belle Glade, FL
	Apr 1971 #06032 farm workers in the field; Belle Glade, FL	Apr 1971 #06041 Belle Glade, FL
	Dec 1971 #09929 harvesting tomatoes; Palmetto, FL	Mar 1971 #04742 harvesting celery; Seminole County, FL
63	Dec 1971 #10042 Dead Eye stopping for a smoke; Palmetto, FL	Apr 1971 #06031 Pompano, FL
	Apr 1971 #06027 Roberta Frasier harvesting tomatoes; Palmetto, FL	Dec 1971 #11125 Reverend Peterson out in the field; Orlando, FL
64	Dec 1971 #11514 farmers with goat in orange field; Lakeland, FL	Dec 1971 #11213 B Farmworker strike of C Cook & Sons Enterprises; Haines City, FL
	Dec 1971 #11206 farm worker day crew selection & early morning pickup; Lakeland, FL	Dec 1971 #11522 Dead Eye and Clyde in the grove
	Apr 1971 #04905 unloading orange harvest at Snow Cap Products; Lake City, FL	May 1971 #06722 farm worker homestead, plowing the fields by horse drawn plow; Laurenburg, NC
65	Dec 1971 #10918 tobacco and rolling papers; Lakeland, FL	
	May 1971 #06719 Isaac McNarr plowing his fields by horse drawn plow; Laurenburg, NC	
68	Dec 1971 #10334 John Wilcox tells his story while in residence with his wife Claudia, son Isaac, and new daughter at his mother-in-law Cat's home; Lakeland, FL	
72	Dec 1971 #10242 caught a possum, yum; Palmetto, FL	Dec 1971 #10240 cooked possum; Palmetto, FL
	Dec 1971 #10337 gutting cooked possum; Palmetto, FL	Dec 1971 #10139 crew hanging out in kitchen; Palmetto, FL
	Dec 1971 #10138 Tom Policano with John Wilcox; Palmetto, FL	Dec 1971 #10028 John Wilcox showing off dinner; Palmetto, FL
73	Dec 1971 #10715 hanging out at mother-in-law Cat's home; Lakeland, FL	Dec 1971 #10716 hanging out at mother-in-law Cat's home; Lakeland, FL
	Dec 1971 #10714 Claudia watching the TV that is always on; Lakeland, FL	Dec 1971 #10717 hanging out at mother-in-law Cat's home; Lakeland, FL
	Dec 1971 #12008 John Wilcox passed out at the end of the day; Lakeland, FL	Dec 1971 #11908 John Wilcox passed out at the end of the day; Lakeland, FL
74	Dec 1971 #11207 Wilcox and some of the crew "waitin on man" to go to the fields; Lakeland, FL	Dec 1971 #11208 Wilcox and some of the crew "waitin on man" to go to the fields; Lakeland, FL
	Dec 1971 #11410 Wilcox helping fix a car on a day with no work; Lakeland, FL	Dec 1971 #11419 Wilcox helping fix a car on a day with no work; Lakeland, FL
	Dec 1971 #11139 Wilcox and in-laws fooling around; Lakeland, FL	Dec 1971 #11718 Xmas time; Lakeland, FL

75	Dec 1971 #11822 records at home; Lakeland, FL	Dec 1971 #12007 Virgin Mary picture at home; Lakeland, FL
	Dec 1971 #11930 Isaac and sister at grandma's; Lakeland, FL	Dec 1971 #12006 portrait on wall; Lakeland, FL
	Dec 1971 #11406 John Wilcox pointing out bike he wants to get son Isaac for xmass; Lakeland, FL	Dec 1971 #12016 John Wilcox; Palmetto, FL
76	Dec 1971 #10505 Cat cooking; Lakeland, FL	Dec 1971 #11221 Cat's kitchen; Lakeland, FL
	Dec 1971 #11217 frying dinner in Cat's kitchen; Lakeland, FL	Dec 1971 #11222 frying dinner in Cat's kitchen; Lakeland, FL
	Dec 1971 #11223 frying chicken in Cat's kitchen; Lakeland, FL	Dec 1971 #10102 Isaac and sister in the kitchen; Palmetto, FL
77	Dec 1971 #11224 Claudia cooking in mother Cat's kitchen; Lakeland, FL	
78	Dec 1971 #11926 Johnny Jr; Lakeland, FL	Dec 1971 #11132 Wilcox out to get milk for Jonny Jr; Lakeland, FL
	Dec 1971 #11830 Liquor store; Lakeland, FL	Dec 1971 #11829 Wilcox buying scotch; Lakeland, FL
	Dec 1971 #11826 Wilcox horsing it up with in-laws; Lakeland, FL	Dec 1971 #11828 Wilcox horsing it up with in-laws; Lakeland, FL
79	Dec 1971 #11820 Wilcox at in-laws; Lakeland, FL	Dec 1971 #11916 Wilcox horsing it up with in-laws; Lakeland, FL
	Dec 1971 #11918 Claudia with cousins and Wilcox; Lakeland, FL	Dec 1971 #11913 Wilcox horsing it up with in-laws; Lakeland, FL
	Dec 1971 #11911 Wilcox horsing it up with in-laws; Lakeland, FL	Dec 1971 #11920 Wilcox horsing it up with in-laws; Lakeland, FL
80	May 1971 #06442 Ben Samuels	Dec 1971 #11604 Luke horsing around for photo
	Mar 1971 #03932 Atlanta, GA	Mar 1971 #04431 Leonard Smith Atlanta, GA
	Mar 1971 #03933 Charley; Atlanta, GA	Dec 1971 #10324 Presscott and Luvinia
81	Apr 1971 #05039 Fort Pierce, FL	Dec 1971 #11701 Charley
	Dec 1971 #11013 Winter Garden, FL	Dec 1971 #10330 Mighty Ark
	Mar 1971 #03928 Tom Policano; Atlanta, GA	Nov 1971 #09724 Mighty Ark
82	Dec 1971 #11403 waiting	Dec 1971 #10517 Lakeland, FL
	Apr 1971 #05033 Fort Pierce, FL	Mar 1971 #04736 Haines City, FL
	Apr 1971 #05037 Fort Pierce, FL	Apr 1971 #06004 Pompano, FL
83	Nov 1971 #09516 Bartow, FL	May 1971 #06407 Frostproof, FL
	Mar 1971 #04735 Haines City, FL	Apr 1971 #05034 Fort Pierce, FL
	Apr 1971 #06042 Belle Glade, FL	Apr 1971 #05738 C Blyther; NC
84	May 1971 #06711 Charleston, SC	Apr 1971 #05027 Fort Pierce, FL
	Dec 1971 #11608 shoes	Apr 1971 #05141 "NO - Niggers, Puertorican, Mexicans *Allowed*"; Wanita, SC
	Dec 1971 #10540 shoes; Lakeland, FL	May 1971 #06122 Pompano, FL

85	Dec 1971 #11219 Lakeland, FL	Dec 1971 #11609 Luke's shoes
	Nov 1971 #09509 Lakeland, FL	Apr 1971 #05140 "Colored" (separate entrance); Wanita, SC
	Oct 1971 #09324 Luke's woman recently buried	Apr 1971 #05935 Indian Town, FL
86	Nov 1971 #09728 Mighty Ark; Palmetto, FL	Dec 1971 #10327 Palmetto, FL
	Dec 1971 #10100 Mighty Ark; Palmetto, FL	
87	Dec 1971 #10326 mamma	Dec 1971 #10539 Lakeland, FL
	Apr 1971 #05741 C Blyther	Dec 1971 #11716 Luke; Palmetto, FL
88	Dec 1971 #11129 Reverand Peterson with his extended family at work in the fields; Orlando, FL	
96	Apr 1971 #05041 workers gathering to march in Tallahassee; Fort Pierce, FL	Dec 1971 #11008 Winter Garden, FL
	Sep 1971 #08825 NYS Legislature nearing onmigrant labor problems; Albany, NY	Nov 1971 #09617 United Farm Workers Organizing Committee; Ochoee, FL
	Apr 1971 #05912 Florida Farm Workers protest bus; Lakeland, FL	Jul 1971 #07114 Jake Paterson director of Peoples Opportunity Incorporated; Albion, NY
97	Jul 1971 #07118 Peoples Opportunity Incorporated; Albion, NY	Nov 1971 #09502 Elmer Dixon & Louis of the Florida Farm Workers Organization; Lakeland, FL
	May 1971 #06610 If It's Good For The Grower Do It; Lakeland, FL	
98	Mar 1971 #04419 Bavard Sentinel news clippings; Orlando, FL	Dec 1971 #11005 farm worker related news clippings; NYS
99	Apr 1971 #06 Migrant And Seasonal Farm Worker Powerlessness congressional hearings testimony; Washington, DC	
100	Apr 1971 #05613 Florida Farm Worker member (FFW); Tallahassee, FL	Apr 1971 #05617 (FFW); Tallahassee, FL
	Apr 1971 #05616 (FFW); Tallahassee, FL	Apr 1971 #05615 (FFW); Tallahassee, FL
	Apr 1971 #05614 (FFW); Tallahassee, FL	Apr 1971 #05612 (FFW); Tallahassee, FL
101	Dec 1971 #12201 Donald Duck Citrus Products Plant; Haines City, FL	Dec 1971 #11002 agricultural chemicals plants; Orlando, FL
	Nov 1971 #09514 Lakeland, FL	May 1971 #06607 Florida Department of Citrus truck; Lakeland, FL
	Mar 1971 #04707 orange grove from atop a fire tower; Haines City, FL	May 1971 #06128 unchecked fire due to drought; Everglades, FL

www.ingramcontent.com/pod-product-compliance
Lightning Source LLC
LaVergne TN
LVHW070954210926
848482LV00002B/213